Around the World in 21 Days

Inventions & Discoveries Version!

Leon Sky

Table of Contents:

Day 1: Canada 7
    The story of the discovery of Insulin 9
    The story of the discovery of Bacteriophages 10

Day 2: Russia 13
    The story of the invention of the Space Station 15
    The story of the innovation of the Periodic Table 18
    The story of the invention of solar cells 21

Day 3: Japan 25
    The story of the invention of the QR Code 27
    The story of the invention of the electronic pocket calculator 28

Day 4: South Korea 31
    The story of the innovation of OLED 32

Day 5: China 35
    The story of the innovation of Paper production 37
    The story of the invention of Silk production 39
    The story of the invention of Acupuncture 41
    The story of the invention of the seismograph 43
    The story of the invention of mechanical clocks 45

Day 6: India 49
    The story of the invention of Zero 51
    The story of the invention of Yoga 54
    The story of the invention of Chess 57
    The story of the invention of Binary system 59

Day 7: Pakistan 61
    The story of the invention of the Computer Virus 62
    The story of the invention of the Ommaya Reservoir 65

Day 8: Iran **69**
    The story of the invention of Sulfuric acid production 71
    The story of the invention of the Qanat system 74
    The story of the innovation of Algebra 76
    The story of the invention of Alcohol extraction 80
    The story of the Gondishapur University 83

Day 9: Turkey **87**
    The story of the innovation of the Vaccination 89

Day 10: Greece **91**
    The story of the invention of Democracy 93
    The story of the invention of the Olympic Games 97
    The story of the invention of the Lighthouse 100

Day 11: Italy **103**
    The story of the invention of Eyeglasses 105
    The story of the invention of the Electric Battery 107
    The story of the invention of the Maser and Laser technologies 111

Day 12: Switzerland **115**
    The story of the invention of the Red Cross 117

Day 13: Germany **121**
    The story of the invention of the Automobile 123
    The story of the discovery of X-ray 126
    The story of the invention of the Jet engine 129

Day 14: Denmark **133**
    The story of the invention of the Bluetooth 135

Day 15: Sweden **139**
    The story of the invention of the Pacemaker 141

Day 16: Norway **144**
    The story of the invention of the Gas turbine 145

Day 17: England **149**
   The story of the invention of the Television.. 151
   The story of the innovation of Vaccination 153
   The story of the discovery of the Penicillin... 155
   The story of the invention of the Steam engine 158

Day 18: France **161**
   The story of the invention of Pasteurization...................................... 163
   The story of the invention of Photography 165
   The story of the invention of the Braille system 167
   The story of the invention of the Stethoscope 170

Day 19: Spain **173**
   The story of the invention of the Autotransfusion device 175

Day 20: Morocco **177**
   The story of the innovative creation of the World Atlas 179

Day 21: USA **183**
   The story of the invention of the Internet 186
   The story of the invention of the Airplane... 189
   The story of the invention of the Polio Vaccine 191
   The story of the invention of the MRI... 194
   The story of the invention of the Email.. 198
   The story of the invention of the ATM 201

Day 1: Canada

Canada has contributed a range of significant innovations, explorations, and inventions that have impacted technology, medicine, engineering, and daily life worldwide. Here are fifteen important contributions made by Canadians:

1. **Insulin**: In 1921, Dr. Frederick Banting and Charles Best discovered insulin at the University of Toronto, revolutionizing diabetes treatment and saving millions of lives globally.

2. **Telephone**: Alexander Graham Bell, who spent much of his career in Canada, invented the first practical telephone in 1876. This innovation transformed global communication and led to the development of modern telecommunications.

3. **IMAX**: Canadian filmmakers Graeme Ferguson, Roman Kroitor, and Robert Kerr developed the IMAX system in the late 1960s. IMAX theaters now provide immersive, high-quality film experiences worldwide.

4. **Pablum**: This nutrient-rich baby cereal was developed in the 1930s at Toronto's Hospital for Sick Children by Drs. Alan Brown, Frederick Tisdall, and Theodore Drake, providing a lifesaving source of nutrition for infants worldwide.

5. **Standard Time Zones**: Sir Sandford Fleming, a Canadian engineer, proposed the concept of worldwide standard time zones in 1878. His system was adopted globally, creating consistency in timekeeping.

6. **Snowmobile**: Joseph-Armand Bombardier invented the first practical snowmobile in 1937, creating a means of transportation in snowy regions and leading to the snowmobile's popularity in recreational and rescue operations.

7. **Basketball**: Dr. James Naismith, a Canadian physical education instructor, invented basketball in 1891 as an indoor sport. It has since become one of the most popular sports globally.

8. **Electron Microscope**: The first practical electron microscope was developed in 1938 by Canadian scientists Eli Franklin Burton and

students at the University of Toronto, allowing for unprecedented magnification of small particles.

9. **Walkie-Talkie**: Canadian engineer Donald Hings developed the first walkie-talkie during World War II for military communication, which later found use in many civilian applications.

10. **The Canadarm**: Canada's robotic arm, developed in the 1980s for the Space Shuttle program, became essential for space missions, assisting with satellite deployment and repairs. It remains a key feature on the International Space Station.

11. **Artificial Cardiac Pacemaker**: Dr. John Hopps, a Canadian electrical engineer, developed one of the first cardiac pacemakers in the 1950s, providing life-saving technology for heart patients worldwide.

12. **The Paint Roller**: Norman Breakey, a Canadian from Toronto, invented the paint roller in the 1940s, which simplified painting jobs and became a staple tool for professionals and DIY enthusiasts alike.

13. **Java Programming Language**: James Gosling, a Canadian computer scientist, created Java in 1994 while working at Sun Microsystems. Java has since become one of the most widely used programming languages.

14. **Superman**: Though a fictional character, Superman was co-created by Canadian artist Joe Shuster along with writer Jerry Siegel. Superman became one of the most iconic superheroes, launching the global superhero genre in comic books, film, and television.

15. **Bacteriophages**: Félix d'Hérelle advanced his groundbreaking discovery of bacteriophages at Université Laval, exploring their potential to treat bacterial infections. His pioneering work on phage therapy laid the foundation for alternatives to antibiotics, particularly against antibiotic-resistant bacteria.

These contributions reflect Canada's impact on diverse fields such as medicine, technology, and popular culture, highlighting the country's innovative spirit.

The story of the discovery of Insulin

The discovery of insulin is one of the most important medical breakthroughs of the 20th century, and it is often attributed to the work of Canadian scientists, particularly Frederick Banting, Charles Best, John Macleod, and James Collip.

In the early 1920s, diabetes was a deadly disease, as there was no effective treatment. Patients with type 1 diabetes, in which the pancreas produces little to no insulin, would eventually die from the condition. At the time, scientists knew that the pancreas was involved in regulating blood sugar, but they didn't fully understand its role or how to treat diabetes.

The Beginning of the Discovery

Frederick Banting, a young surgeon from Ontario, had an idea that insulin might be the key to treating diabetes. He theorized that the pancreas contained a substance that could regulate blood sugar, and he believed this substance was produced in small clusters of cells within the pancreas, called the islets of Langerhans. Banting proposed that if these cells could be isolated and the substance extracted, it could help people with diabetes.

In 1921, Banting approached John Macleod, a professor of physiology at the University of Toronto, with his idea. Macleod, who was skeptical at first, gave Banting access to his lab and also suggested that Banting work with a medical student, Charles Best, to conduct the experiments.

The Key Experiments

Banting and Best began their experiments by tying off the ducts of the pancreas in dogs, which caused the pancreas to stop producing digestive enzymes while allowing the insulin-producing islets of Langerhans to remain intact. They then extracted a substance from the pancreas and tested it on diabetic dogs. Their initial experiments showed promising results: the dogs' blood sugar levels dropped after being injected with the extracted substance.

At first, Banting and Best struggled to isolate enough of the substance to treat human patients. It was only after the addition of James Collip, a biochemist, to their team that they were able to purify the extract to a level where it could be used effectively in humans.

The First Successful Treatment

In January 1922, the team successfully tested insulin on a 14-year-old diabetic patient named Leonard Thompson at Toronto General Hospital. The treatment was a success, and Thompson's blood sugar levels dropped significantly. This breakthrough marked the first time that insulin had been used to treat diabetes, offering hope to millions of people suffering from the disease.

***

Banting, Best, Macleod, and Collip were quickly recognized for their groundbreaking work. In 1923, Banting and Macleod were awarded the Nobel Prize in Physiology or Medicine for their discovery of insulin. Banting, however, shared his prize money with Charles Best, acknowledging his important contribution to the work.

The discovery of insulin revolutionized the treatment of diabetes, turning it from a fatal disease into a manageable condition. Insulin therapy became widely available, and millions of lives have been saved since.

Banting's discovery is still celebrated today, not only for its immediate impact on public health but also for its example of scientific collaboration and perseverance. The story of insulin's discovery remains one of the most remarkable achievements in the history of medicine.

The story of the discovery of Bacteriophages

The exploration of bacteriophages, or "phages," by Félix d'Hérelle is a landmark story in microbiology, marked by a discovery that opened new doors in fighting bacterial infections. Although originally from France, d'Hérelle's exploration of bacteriophages took on new dimensions in Canada, where he conducted some of his most influential research.

1. Early Interest and Discovery of Bacteriophages

In 1917, while working at the Pasteur Institute in Paris, d'Hérelle observed a mysterious phenomenon: a substance in filtered samples from patients with dysentery seemed to destroy bacterial cultures. After investigating further, he realized he had discovered a new type of virus that specifically targeted and killed bacteria. D'Hérelle named these viruses "bacteriophages," meaning "bacteria eaters."

2. Arrival in Canada and Research at Université Laval

In the 1920s, d'Hérelle moved to Canada, joining Université Laval in Quebec. Canada offered a new scientific environment where he could further explore bacteriophages and their therapeutic potential. At Université Laval, he worked to purify phages and test their effectiveness against various bacterial infections. This research allowed d'Hérelle to establish Canada as an early center for bacteriophage research, especially as antibiotic treatments were not yet available.

3. Testing Phage Therapy

D'Hérelle was an early proponent of using phages as an alternative to antibiotics, a practice known as *phage therapy*. In Canada, he conducted experiments and clinical trials to assess the safety and efficacy of phage therapy for treating bacterial infections such as dysentery and typhoid fever. His studies showed promising results, demonstrating the potential of phages to target and eliminate bacterial pathogens in patients.

4. Challenges and Controversy

Despite promising findings, d'Hérelle faced challenges in convincing the broader medical community of the value of phage therapy. The lack of standardized protocols and the variability of phage effectiveness led to skepticism. However, d'Hérelle's research in Canada contributed significantly to the development of phage science and inspired future generations of researchers.

5. Legacy of D'Hérelle's Work in Canada

Félix d'Hérelle's work in Canada established a legacy in the field of bacteriophages and laid the groundwork for future exploration into phage therapy, which has seen a resurgence of interest in recent years due to antibiotic resistance. His discoveries are now considered foundational, and his pioneering work in both France and Canada cemented his status as a visionary in microbiology.

D'Hérelle's exploration of bacteriophages in Canada played a key role in advancing the understanding of phage biology and their potential as a therapeutic tool, creating a lasting impact on microbiology and modern medicine.

Day 2: Russia

Russia has made significant contributions to science, technology, arts, and exploration, influencing a wide range of fields. Here are fifteen important innovations, explorations, and inventions introduced by Russians to the world:

1. **Space Exploration (First Satellite - Sputnik)**: In 1957, the Soviet Union launched *Sputnik 1*, the world's first artificial satellite, marking the beginning of the Space Age and paving the way for global space exploration.

2. **Manned Space Flight**: In 1961, cosmonaut Yuri Gagarin became the first human to orbit Earth aboard *Vostok 1*. This achievement showcased the potential for human space exploration and inspired future space missions.

3. **Tetris**: Created by Russian computer engineer Alexey Pajitnov in 1984, *Tetris* became one of the most popular and enduring video games, impacting the gaming industry and influencing puzzle games worldwide.

4. **Periodic Table**: Dmitri Mendeleev, a Russian chemist, developed the Periodic Table of Elements in 1869. His table organized elements by atomic weight and predicted properties of undiscovered elements, forming the basis of modern chemistry.

5. **Radio (Theory of Electromagnetic Waves)**: Alexander Popov, a Russian physicist, made significant advancements in the early development of radio by demonstrating wireless communication in 1895. He is often credited as one of the inventors of radio.

6. **Helicopter Design**: Igor Sikorsky, a Russian-American aviation pioneer, developed the first practical helicopter in the 1930s, creating a model that influenced modern helicopter design and is widely used in aviation today.

7. **AK-47**: Developed by Mikhail Kalashnikov in 1947, the AK-47 is one of the most iconic and widely used assault rifles worldwide due to its durability, simplicity, and reliability in various conditions.

8. **Solar Cells**: Russian scientist Aleksandr Stoletov created the first practical solar cell in the 1880s, demonstrating the photoelectric effect. This discovery laid the groundwork for modern solar energy technology.

9. **Electrification (GOELRO Plan)**: The Soviet Union implemented the GOELRO Plan in the early 20th century, which was one of the world's first large-scale electrification projects. It inspired similar infrastructure projects in other countries and set the stage for modern energy networks.

10. **Theremin (Electronic Musical Instrument)**: Invented by Leon Theremin in the 1920s, the theremin is one of the first electronic musical instruments. It's played without physical contact and influenced the development of electronic music.

11. **Space Station (Mir)**: The Soviet Union launched *Mir*, the first modular space station, in 1986. It served as a scientific research platform in space for over 15 years, pioneering many techniques used in today's International Space Station (ISS).

12. **Non-Euclidean Geometry**: Russian mathematician Nikolai Lobachevsky developed non-Euclidean geometry in the 19th century, challenging classical Euclidean geometry and influencing fields like physics, astronomy, and modern geometry.

13. **Kremlin and Russian Architectural Styles**: Russian architectural innovations, including the onion-dome structures seen in St. Basil's Cathedral and the Kremlin, are world-renowned. These designs reflect Russian engineering, cultural, and artistic heritage.

14. **Katyusha Rocket Launcher**: Developed during World War II, the Katyusha was a highly effective mobile rocket launcher. Its design influenced future military rocket systems and became a symbol of Soviet military innovation.

15. **Hypersonic Technology**: In recent years, Russian advancements in hypersonic missile technology, such as the Avangard hypersonic glide vehicle, have introduced new possibilities for high-speed, long-range defense systems.

These Russian innovations reflect a strong legacy in science, engineering, and culture, with lasting global impacts in fields ranging from space exploration and chemistry to military technology and the arts.

The story of the invention of the Space Station

The Mir Space Station was a groundbreaking achievement by the Soviet Union and later Russia, marking the first modular space station to be assembled in orbit. It was a landmark in space exploration, setting the stage for international collaboration and technological advancements that would influence future space stations, including the International Space Station (ISS).

Origins and Vision for Mir

In the 1970s, the Soviet Union already had substantial experience in space, having launched the Salyut series of space stations starting in 1971. These early Salyut stations were largely one-piece structures, limiting their potential for long-term missions and scientific experiments. As space technology progressed, Soviet engineers envisioned a more advanced station—one that could be expanded over time by adding new modules with different capabilities. This idea became the concept for Mir, which means "peace" or "world" in Russian.

Designing Mir: The Modular Space Station

Mir was designed as a modular station, which meant that it would be assembled in space over time from multiple segments. This concept allowed the station to evolve, with new modules added as the mission expanded or as technology advanced. This approach was revolutionary because, unlike previous single-launch stations, Mir could be tailored and extended for various research needs.

The central core module, known as the base block, would provide basic life-support systems, crew quarters, and docking ports for additional modules. Each subsequent module would serve a unique purpose, from scientific research to power generation and storage.

Launch and Assembly of Mir

1. **Launch of the Core Module**: The first piece of Mir, the Core Module, was launched on February 19, 1986, from the Baikonur Cosmodrome. This module provided essential life-support

systems, control, communication, and living quarters for the crew. After reaching orbit, the Core Module became the foundation around which the entire Mir space station would be built.

2. **Expansion with Specialized Modules**: Over the next decade, additional modules were launched and added to the Core Module, including:

    - **Kvant-1** (1987): This astrophysics module housed instruments for observing cosmic phenomena, X-rays, and gamma rays.
    - **Kvant-2** (1989): Equipped with systems for life support, water recycling, and improved power supplies. It also had an airlock, which allowed for spacewalks.
    - **Kristall** (1990): A laboratory for materials science experiments and biological research. It also contained a docking port for the Soviet shuttle Buran (though Buran was never launched to Mir).
    - **Spektr** (1995): Designed for Earth and atmospheric observations. It was initially planned as a military module but was converted to support scientific research.
    - **Priroda** (1996): The last major module, dedicated to Earth observation and environmental monitoring.

These modules were attached and integrated in orbit, each adding new capabilities and expanding the station's research potential.

Life on Board Mir

Living and working on Mir was challenging but rewarding. The station hosted dozens of missions, with cosmonauts and astronauts spending months at a time on board, performing scientific research, conducting experiments, and testing the effects of long-term space exposure on the human body.

In 1987, Yury Romanenko set a world record by spending 326 continuous days aboard Mir, providing invaluable insights into the physical and psychological effects of extended space habitation. Over its lifetime, Mir hosted crew members from multiple countries, including the United

States, Japan, Germany, and France, fostering a spirit of international collaboration.

International Collaboration: The Shuttle-Mir Program

Following the end of the Cold War, Mir became a platform for collaborative missions between Russia and the United States. In 1993, the two countries launched the Shuttle-Mir Program, in which NASA's Space Shuttle would dock with Mir and transport crew and supplies. This partnership helped repair tensions between the two former Cold War adversaries and allowed NASA to gain valuable experience in long-duration space missions, a crucial step toward developing the International Space Station (ISS).

The Shuttle-Mir Program also enabled Russian and American scientists to work together, merging research programs and sharing data. It was the first large-scale cooperative space program between the two nations and helped lay the groundwork for future international space partnerships.

Mir's Legacy and Deorbit

By the late 1990s, after more than a decade in orbit, Mir began showing signs of wear. The station experienced technical issues, including power failures, equipment breakdowns, and even a fire in 1997. Although Russian engineers and cosmonauts worked to keep Mir operational, the cost of maintaining the aging station was substantial, and resources were increasingly being directed toward the ISS.

In 2001, after 15 years of groundbreaking work, Mir was intentionally deorbited. It re-entered Earth's atmosphere on March 23, 2001, over the South Pacific Ocean, breaking apart in the atmosphere and bringing an end to its historic mission.

Mir's Contributions to Space Exploration

Mir was the first truly modular space station and the longest continuously inhabited spacecraft until that time. It taught engineers and scientists crucial lessons about long-term space operations, such as:

- **Crewed Space Habitation**: Mir allowed scientists to study the effects of prolonged spaceflight on the human body, knowledge that is essential for future missions to Mars and beyond.
- **Modular Space Station Design**: Mir's modular design proved that space stations could be built and expanded over time, a principle used in the ISS.
- **International Collaboration**: Mir served as a stepping stone for global partnerships in space, particularly through the Shuttle-Mir Program, paving the way for the ISS.
- **Operational Challenges in Space**: The station's operational issues highlighted the challenges of maintaining complex systems in space, leading to improvements in station design and maintenance for future missions.

Mir remains a symbol of technological achievement and human perseverance. It was not only a triumph for Russian engineering but also a critical step in advancing international space exploration. Its legacy lives on in the International Space Station and in all future space missions that rely on the pioneering lessons learned from the Mir space station.

The story of the innovation of the Periodic Table

The story of the innovation of the Periodic Table begins in Russia with the pioneering work of Dmitri Mendeleev, a Russian chemist who is widely credited with the creation of the modern Periodic Table of Elements. Mendeleev's work in the 19th century revolutionized the way scientists understood the elements and their relationships to one another.

Early Efforts to Classify Elements

Before Mendeleev's breakthrough, scientists knew of many different chemical elements, but there was no systematic way to organize them. In the early 19th century, elements were typically classified based on their physical and chemical properties, but there was no clear understanding of how they were related or why they exhibited similar behaviors.

Several chemists had attempted to organize the elements. For example, in the 1820s, Johann Döbereiner, a German chemist, discovered what came to be known as Döbereiner's Triads, where he noticed that groups of three elements had similar properties, and the atomic weight of the middle element was approximately the average of the other two. However, this was not a comprehensive system and didn't explain all elements.

Other chemists, such as John Newlands in 1864, also attempted to classify elements by observing periodic patterns, leading to the concept of the law of octaves, where every eighth element seemed to show similar properties. However, these early attempts were also incomplete.

Mendeleev's Breakthrough: The Periodic Law

In the early 1860s, Dmitri Mendeleev was working as a professor of chemistry at the University of St. Petersburg in Russia. He was tasked with writing a textbook on chemistry, and in doing so, he became frustrated with the lack of a logical way to organize the elements. Mendeleev began experimenting with different methods of arranging the elements according to their atomic masses.

In 1869, Mendeleev's breakthrough came when he organized the known elements into a table based on their atomic masses. What Mendeleev noticed was that when the elements were arranged in order of increasing atomic mass, certain elements exhibited recurring patterns of chemical and physical properties. This led him to the realization that the properties of elements were periodic functions of their atomic masses, which became known as the Periodic Law.

The Structure of the Periodic Table

Mendeleev's table was remarkable because it showed that elements with similar properties appeared at regular intervals. He arranged the elements in rows (periods) and columns (groups), and when he left spaces for elements that had not yet been discovered, he correctly predicted their properties. For example, Mendeleev predicted the existence and properties of gallium (then known as eka-aluminum), germanium (eka-silicon), and scandium (eka-boron), based on the gaps in his table. When these elements were later discovered, they matched Mendeleev's predictions almost exactly.

One of the most significant features of Mendeleev's periodic table was his ability to predict the existence of undiscovered elements. This gave Mendeleev's table credibility and helped secure his place in history as the father of the modern Periodic Table.

The Recognition and Impact

Mendeleev's table, first published in 1869, had an immediate impact on the scientific community. Initially, there were some skeptics, and some chemists criticized Mendeleev for arranging the elements by atomic mass rather than atomic number. However, over time, his system gained widespread acceptance, and it was eventually found that Mendeleev's arrangement was more accurate than earlier systems.

In the years following the publication of the Periodic Table, scientists discovered more elements and refined the structure of the table. The understanding of atomic number, which was introduced by Henry Moseley in 1913, eventually replaced atomic mass as the organizing principle. Moseley's work showed that the elements fit better into the table when ordered by atomic number rather than atomic mass, resolving some of the inconsistencies in Mendeleev's original arrangement.

Mendeleev's Legacy

Mendeleev's innovation fundamentally changed the way chemists understood the relationships between elements and laid the groundwork for modern chemistry. The Periodic Table has become one of the most important tools in science, organizing the elements based on their atomic structure and properties. Mendeleev's Periodic Table was one of the first frameworks to show that the properties of elements are not random but are instead influenced by their position in the table.

His work earned him worldwide recognition, and Mendeleev is often referred to as one of the founders of modern chemistry. In Russia, he is celebrated as a national hero, and his legacy continues to shape the field of chemistry to this day.

The innovation of the Periodic Table in Russia by Dmitri Mendeleev was a groundbreaking achievement that transformed the field of chemistry. Mendeleev's careful organization of the elements led to the discovery of patterns in their properties and predicted the existence of new elements. His work not only advanced our understanding of the elements but also laid the foundation for future scientific discoveries in chemistry and physics. The Periodic Table remains an essential tool in science, and Mendeleev's legacy continues to inspire chemists around the world.

The story of the invention of solar cells

The story of the invention of solar cells in Russia begins with Aleksandr Stoletov, a Russian physicist and inventor who is often credited with laying the groundwork for modern solar cell technology. Stoletov's work in the late 19th century contributed significantly to the understanding of the photoelectric effect, a phenomenon that is foundational to the operation of solar cells.

The Discovery of the Photoelectric Effect

The photoelectric effect refers to the phenomenon in which electrons are ejected from a material (usually metal) when it absorbs light or electromagnetic radiation. The concept of the photoelectric effect was first observed by Heinrich Hertz in 1887, but it was Stoletov's experiments that provided more crucial insights into how light could be used to generate electrical currents.

Aleksandr Stoletov's Contributions:

In 1888, Stoletov began his experiments with the photoelectric effect. He focused on the interaction between light and metals, particularly the way light could be used to generate an electric current. He observed that when light fell on a metal surface, it caused the metal to emit electrons, producing a flow of electric current.

Stoletov's pioneering work in the field of photoelectricity led him to create the first solar cell-like device. In one of his most important experiments, Stoletov placed a metallic plate in a vacuum tube and exposed it to light. He noticed that the intensity of the electric current flowing through the device increased as the intensity of light increased. This was the first demonstration of how light could be converted into electrical energy, laying the foundation for solar power.

Stoletov's work in photoelectricity was one of the earliest studies of what would eventually become the solar cell. His discovery revealed the potential of using light energy to generate electricity, which would later form the basis for the development of solar cells.

The Impact of Stoletov's Work

Stoletov's experiments were groundbreaking for several reasons:

- They established the relationship between light and electricity, a crucial principle for the later development of photovoltaic technology.

- His work demonstrated that metals could emit electrons when exposed to light, a phenomenon that would become critical in the development of solar cells.

- Stoletov's findings directly contributed to the understanding of the photoelectric effect, which was later used by other scientists, including Albert Einstein, who received the Nobel Prize in Physics in 1921 for his work on the photoelectric effect. While Einstein expanded on Stoletov's ideas, Stoletov's earlier experiments were instrumental in providing a foundation for this breakthrough.

The Invention of the Solar Cell

While Stoletov's work did not directly lead to the invention of modern solar cells, it laid the theoretical and experimental groundwork that would later be used by other scientists. In the 1950s, the first practical photovoltaic solar cell was developed by scientists at Bell Labs in the United States, with the work of Charles Fritts and others contributing to the development of early solar cell technology.

Fritts' solar cells, made from selenium and coated with a thin layer of gold, were the first devices capable of converting sunlight into electrical energy with measurable efficiency. However, it was Stoletov's work on the photoelectric effect that helped establish the scientific principles behind the operation of such devices.

***

Aleksandr Stoletov's work in photoelectricity remains a significant milestone in the history of solar energy. His research provided the first insights into how light could be harnessed to generate electricity, paving the way for the eventual development of solar power technologies.

Although Stoletov's name is not as widely recognized in the field of solar energy as some of his contemporaries, his contribution to the understanding of the photoelectric effect has had a lasting impact. His pioneering work is seen as an early precursor to the field of photovoltaic energy, and in Russia, he is still honored as one of the foundational figures in the development of solar technology.

The story of the invention of solar cells in Russia is intricately linked to the research and discoveries made by Aleksandr Stoletov. His work on the photoelectric effect in the late 19th century laid the foundation for the later development of solar cells, marking him as a precursor to the modern field of solar energy. While Stoletov's experiments did not immediately lead to the creation of the solar cell as we know it today, his early work on converting light into electricity remains a key milestone in the history of renewable energy.

Day 3: Japan

Japan has made numerous contributions to technology, medicine, the arts, and everyday life, creating innovations that have impacted the world in transformative ways. Here are fifteen important innovations, explorations, and inventions introduced by Japan:

1. **The Bullet Train (Shinkansen)**: Japan introduced the world's first high-speed rail system, the Shinkansen, in 1964. This innovation redefined public transportation, inspired high-speed rail development worldwide, and showcased the potential of fast, efficient, and safe train travel.

2. **Instant Noodles**: Invented by Momofuku Ando in 1958, instant noodles became a revolutionary food product due to their convenience and affordability. Instant ramen has since become a global staple, popular across cultures and demographics.

3. **QR Code**: The Quick Response (QR) code was developed in 1994 by Masahiro Hara and his team at Denso Wave. QR codes are now widely used for mobile payments, inventory management, and information sharing, becoming especially important in the digital age.

4. **Anime and Manga**: Japan's anime and manga industry has had a huge cultural influence globally, creating a distinct art form and storytelling style that has resonated with audiences of all ages and inspired countless creators worldwide.

5. **Digital Cameras**: Japanese companies like Sony, Canon, and Nikon pioneered digital camera technology in the late 20th century. Their innovations transformed photography from film to digital, making it more accessible and shaping modern media.

6. **Sony Walkman**: Launched in 1979, the Sony Walkman was the first portable personal stereo, revolutionizing how people listened to music and paving the way for portable media devices such as MP3 players and smartphones.

7. **Robotics**: Japan is a world leader in robotics. Humanoid robots, such as ASIMO by Honda, set benchmarks in robotics research. Japanese robots are used in fields ranging from manufacturing to

healthcare, helping to redefine the roles of machines in human society.

8. **Toyota Production System (Lean Manufacturing)**: Toyota developed a manufacturing system that focuses on efficiency, quality, and waste reduction. This "lean manufacturing" model has since been adopted by companies worldwide, becoming the standard for efficient production.

9. **Hybrid Cars**: Toyota's Prius, launched in 1997, was the world's first mass-produced hybrid vehicle. This innovation popularized eco-friendly automotive technology, influencing the global shift toward sustainable vehicles.

10. **Video Game Consoles**: Companies like Nintendo, Sony, and Sega played pivotal roles in popularizing video game consoles, starting with devices like the NES (Nintendo Entertainment System) and later the PlayStation. These consoles reshaped global entertainment and gaming culture.

11. **Blue LED Technology**: Professors Isamu Akasaki, Hiroshi Amano, and Shuji Nakamura developed the blue LED in the 1990s, enabling white LED lighting. This invention earned a Nobel Prize and contributed to energy-efficient lighting systems worldwide.

12. **Electronic Pocket Calculator**: In the early 1970s, Japanese companies such as Casio and Sharp created the first compact electronic calculators, making complex calculations more accessible and revolutionizing education and business practices.

13. **Capsule Hotels**: Introduced in 1979, capsule hotels provide compact, efficient accommodations in crowded urban areas. This concept has since spread globally as a unique, affordable lodging option.

14. **Emoji**: Shigetaka Kurita developed the first set of emoji in the late 1990s while working for NTT DoCoMo. These small digital images have become a universal language in online communication, used to convey emotions and ideas visually across languages and cultures.

15. **JIT (Just-in-Time) Inventory System**: Pioneered by Toyota, the Just-in-Time inventory system transformed supply chain

management, reducing waste and improving efficiency. This system is now used worldwide in various industries.

These contributions from Japan have left lasting impacts across fields, showcasing Japan's strengths in innovation, efficiency, and creativity, and reshaping global technology, culture, and lifestyle.

The story of the invention of the QR Code

The QR Code (Quick Response Code) was invented in Japan in 1994 by a Japanese company called Denso Wave, a subsidiary of the Toyota Group. The QR Code was developed as a more efficient way to track vehicles during the manufacturing process.

The story begins with the need for a barcode system that could store more information and allow faster scanning. Traditional barcodes could only hold a limited amount of data and were slow to scan, especially when used in a fast-paced production environment like car manufacturing.

Denso Wave's engineer, Masahiro Hara, led the development of the QR Code. He designed it to store a significant amount of data, be scannable from any direction, and have fast scanning capabilities. This was achieved by using a two-dimensional (2D) grid instead of a one-dimensional line system, allowing it to hold much more information—up to 7,000 characters—compared to traditional barcodes, which were limited to about 20 characters.

The QR Code also included a feature called error correction, which allowed the code to remain scannable even if parts of it were damaged. Its black-and-white square design with distinctive corner markers made it easier to read quickly, regardless of the angle or orientation.

The QR Code quickly became popular in Japan, where it was used for various purposes, including advertising, ticketing, and product tracking. Its use spread globally, especially with the rise of smartphones with built-in QR code readers, and it became a universal tool for linking physical objects to digital content.

Today, QR codes are used in countless applications around the world, from payments and marketing to event ticketing and logistics.

The story of the invention of the electronic pocket calculator

The invention of the electronic pocket calculator is a significant achievement in the history of technology, and Japan played a key role in its development. The story of this invention involves several key innovations and contributions from Japanese companies and engineers, who revolutionized the way we perform mathematical calculations.

Early Calculating Machines

Before the invention of the electronic pocket calculator, mechanical calculators were used for arithmetic operations. These were large, often cumbersome devices that required manual operation. In the 1960s, the world was transitioning from mechanical to electronic technology, and companies were beginning to explore the possibilities of compact, portable electronic devices that could handle arithmetic calculations.

The First Japanese Electronic Calculators

In the early 1960s, Japanese companies, including Sharp and Casio, began exploring the development of electronic calculators. These early calculators were bulky and expensive, but they marked the beginning of a new era in computing technology.

- **Sharp** Corporation made early strides in developing electronic calculators. In 1964, Sharp introduced one of the first electronic calculators, the Sharp CS-10A. This was a large machine, but it demonstrated the potential for electronic devices to perform calculations.
- **Casio** was another key player in the development of calculators. Casio had already been producing mechanical calculators in the 1950s but began developing electronic models in the 1960s.

The Invention of the Pocket-Sized Calculator

The first true electronic pocket calculator came to fruition in the early 1970s. The breakthrough was made possible by advances in integrated circuits (ICs) and semiconductor technology, which allowed for miniaturization of electronic components.

In 1970, the Japanese company Casio introduced the Casio Mini, the world's first pocket-sized electronic calculator. This innovation was based on the use of a single integrated circuit, which made the device much smaller, more affordable, and more practical for everyday use.

The Key Innovation: Integrated Circuit (IC)

The success of the pocket calculator was largely due to the development of the integrated circuit. Before ICs, electronic components like transistors were large and bulky. The integrated circuit allowed for the combination of many electronic components (such as resistors, transistors, and capacitors) onto a single microchip. This breakthrough made it possible to create small, efficient, and reliable calculators that could fit in your pocket.

Casio, along with other Japanese companies like Sharp, Canon, and Hewlett-Packard, used these integrated circuits to develop and produce calculators that were affordable and accessible to the public.

Commercial Success and Mass Production

The pocket calculator quickly gained popularity in the early 1970s, especially in the fields of business, education, and engineering. As technology improved, the price of these devices dropped, making them accessible to a wider audience. Companies like Texas Instruments and Hewlett-Packard also entered the market, further accelerating the development and adoption of electronic calculators.

Casio's introduction of the Casio fx-1 in 1974 was one of the first truly portable, affordable calculators that could perform basic arithmetic and advanced mathematical functions. These calculators used solar power in later models, making them even more convenient and environmentally friendly.

The Impact on Society and Technology

The introduction of the pocket calculator revolutionized many industries. It greatly improved productivity in areas such as accounting, engineering, science, and education by allowing quick and accurate calculations in a portable format. It replaced the need for large mechanical calculating machines and significantly reduced the time and cost associated with mathematical computations.

The development of the electronic pocket calculator also paved the way for future innovations in personal computing. The principles of miniaturization and affordable electronics that were first applied to calculators would later influence the development of personal computers and smartphones, changing the way people interacted with technology on a daily basis.

***

The invention of the electronic pocket calculator was a pivotal moment in the history of electronics, and Japan played a central role in its development. Companies like Casio, Sharp, and Canon utilized advances in integrated circuits and semiconductor technology to create affordable, portable calculators that transformed industries and everyday life. The invention of the electronic pocket calculator set the stage for the development of modern personal computing devices, making it one of the most influential technological innovations of the 20th century.

Day 4: South Korea

South Korea has made remarkable contributions to technology, science, and culture, establishing itself as a leader in innovation. Here are ten of the most important innovations, explorations, and inventions introduced by South Koreans:

1. **Samsung's Mobile Technology**: Samsung, South Korea's tech giant, has pioneered advancements in mobile technology, particularly with the Galaxy smartphone line. Samsung's innovations in displays, memory chips, and mobile processors have heavily influenced the global smartphone industry.

2. **World's First CDMA (Code-Division Multiple Access) Network**: South Korea was the first country to adopt CDMA technology for mobile communication in the mid-1990s. This advancement accelerated mobile communication technology and paved the way for 3G and 4G networks worldwide.

3. **OLED Displays**: South Korean companies like LG and Samsung were instrumental in developing and popularizing OLED (Organic Light Emitting Diode) displays. These screens offer superior color accuracy and flexibility, becoming essential in smartphones, TVs, and wearables.

4. **English-Korean Online Dictionary (Naver)**: Naver, South Korea's top search engine, launched one of the world's first comprehensive online dictionaries with multilingual features, including English-Korean. This innovation made online language tools mainstream, inspiring similar resources globally.

5. **Electric and Hydrogen Fuel Cell Vehicles**: South Korean automakers like Hyundai have been at the forefront of eco-friendly vehicle technology, producing both electric and hydrogen fuel cell vehicles. Hyundai's Nexo, for example, is one of the first mass-produced hydrogen-powered vehicles, promoting sustainable transportation.

6. **Video Games and eSports**: South Korea has greatly influenced the global gaming and eSports industry, being the birthplace of competitive online gaming. Games like *StarCraft* and *League of*

Legends popularized eSports, and South Korea's infrastructure has supported the growth of professional gaming worldwide.

7. **Artificial Intelligence and 5G Technology**: South Korea was among the first countries to roll out a nationwide 5G network, setting standards for global telecommunications and internet speeds. South Korean companies also lead in AI research, implementing AI in everything from customer service to smart home technology.

8. **Advanced Cosmetic Products and Skincare (K-Beauty)**: The K-beauty industry, known for innovative skincare products like BB cream, sheet masks, and snail mucin treatments, has influenced global beauty standards and practices. South Korean beauty products are now widely used and emulated worldwide.

These innovations reflect South Korea's strengths in technology, culture, and lifestyle, positioning the country as a global influencer across diverse fields.

The story of the innovation of OLED

The invention of OLED (Organic Light Emitting Diode) technology can be traced back to research conducted in the United States in the 1980s, but South Korea played a crucial role in its commercial development, especially in the 2000s.

Early Development:

The roots of OLED technology go back to the discovery of organic materials that could emit light when an electric current passed through them. In 1987, researchers at Eastman Kodak in the U.S. made a breakthrough by developing the first practical OLED device. This was followed by further advancements in the field, particularly by Ching W. Tang and Steven Van Slyke at Kodak, who created a high-efficiency OLED that could emit light from organic compounds. Their work laid the foundation for future OLED development.

South Korea's Role:

While OLED technology was initially researched in the U.S., South Korea became a key player in its commercialization, especially in the 2000s. The country's technology giants, Samsung and LG Electronics, made significant strides in adapting OLEDs for display technologies, particularly for smartphones, televisions, and flexible displays.

1. **Samsung's Role**: Samsung was a pioneer in bringing OLED to mass markets. In the mid-2000s, Samsung started developing OLED screens for mobile phones. In 2007, they showcased their first commercial OLED display at the Consumer Electronics Show (CES). Samsung's early commitment to OLED technology enabled them to develop more advanced, larger, and brighter displays for smartphones, and they became one of the first companies to mass-produce OLED displays for mobile devices, including the Samsung Galaxy series.

2. **LG's Contribution**: Around the same time, LG Electronics focused on OLED for televisions. LG developed a unique form of OLED technology known as WOLED (White OLED), which differed from Samsung's approach but also achieved impressive results. By the early 2010s, LG had created ultra-thin, high-quality OLED TVs, positioning itself as a leader in the premium television market. LG's OLED TVs gained significant attention for their exceptional picture quality, deep blacks, and vibrant colors, setting them apart from traditional LED/LCD TVs.

3. **Commercialization and Mass Production**: South Korean companies, especially Samsung and LG, invested heavily in OLED R&D and manufacturing infrastructure, which allowed them to scale up production and drive down costs. By the early 2010s, OLED displays began appearing in a wide range of consumer electronics, from smartphones to high-end televisions, and they quickly became the gold standard for display technology.

South Korea's OLED Leadership:

Today, South Korea remains at the forefront of OLED innovation. Samsung continues to dominate the market for OLED screens used in smartphones, while LG has become a leader in OLED TVs. South Korean companies also continue to lead in the development of flexible OLEDs, used in cutting-edge products like foldable smartphones (e.g., Samsung Galaxy Z Fold).

Key Milestones in OLED Technology by South Korean Companies:

- **2007**: Samsung introduces its first OLED display for mobile phones.
- **2013**: LG launches the first OLED TV on the market.
- **2019**: Samsung launches the first foldable OLED smartphone (Galaxy Fold), marking a major milestone for flexible OLED technology.

In summary, while OLED technology was invented outside of South Korea, it was the country's major tech firms, Samsung and LG, that drove its commercialization and popularization on a global scale. Their innovations in OLED displays transformed industries ranging from mobile phones to televisions, cementing South Korea as a world leader in display technology.

Day 5: China

China has a long history of contributing transformative innovations, inventions, and explorations that have shaped both ancient and modern civilizations. Here are fifteen of the most important contributions introduced by China to the world:

1. **Paper**: Invented around 105 AD during the Han Dynasty, Chinese papermaking revolutionized the way information was recorded and disseminated. Paper production spread across Asia and Europe, becoming a cornerstone for literacy and literature.

2. **Gunpowder**: Developed during the Tang Dynasty (9th century), gunpowder was initially used in fireworks and later adapted for military applications, profoundly altering warfare across the globe.

3. **The Compass**: The earliest compass, invented during the Han Dynasty (2nd century BC), was used for geomancy and navigation. By the Song Dynasty (11th century), it enabled exploration and sea navigation, leading to the Age of Exploration.

4. **Printing (Woodblock and Movable Type)**: Woodblock printing emerged in the Tang Dynasty, followed by movable type printing in the Song Dynasty, allowing books to be printed quickly. This innovation spread to Europe, laying the groundwork for the printing press and the mass dissemination of knowledge.

5. **Silk Production**: Chinese silk was one of the most sought-after commodities along the Silk Road, spurring international trade. The silk-making process was a closely guarded secret for centuries, and its production had lasting cultural and economic impacts.

6. **Porcelain**: Perfected during the Tang and Song dynasties, Chinese porcelain became highly prized around the world for its quality, beauty, and durability, influencing ceramic industries across Europe and Asia.

7. **Tea Cultivation and Brewing**: China pioneered the cultivation and consumption of tea over 4,000 years ago. Chinese tea culture has inspired traditions globally, becoming a foundational beverage in countries from Japan to Britain.

8. **Acupuncture and Traditional Chinese Medicine (TCM)**: Originating over 2,000 years ago, TCM, including acupuncture, herbal medicine, and practices like Tai Chi, is now widely recognized for its therapeutic benefits and has influenced alternative medicine worldwide.

9. **Seismograph**: Invented by Zhang Heng in 132 AD, the world's first seismograph could detect earthquakes and their direction, showcasing ancient China's understanding of geoscience and early disaster management.

10. **The Great Wall and Architectural Techniques**: The Great Wall of China, built and enhanced over several dynasties, showcases advanced ancient engineering and has influenced defensive architecture worldwide. Its construction methods informed large-scale architecture and fortifications across different cultures.

11. **Mechanical Clock**: In 725 AD, Yi Xing and Liang Lingzan developed one of the earliest mechanical clocks, an intricate water-driven device. Later, the famous clock by Su Song (1092) advanced timekeeping technology, inspiring further development in medieval Europe.

12. **Paper Money**: First issued during the Tang Dynasty, paper currency became widely used during the Song Dynasty. It allowed for efficient, standardized trade and greatly influenced the development of modern monetary systems.

13. **Kite**: Kites were invented in China around 2,000 years ago and used for various purposes, including military signaling. Kiting eventually became a popular recreational activity worldwide and contributed to early aerodynamics research.

14. **The Abacus**: Known as the "Suanpan," the abacus was developed as a calculating tool in ancient China, and its use spread to other parts of Asia. It became a foundational educational tool for arithmetic and influenced early mechanical calculators.

15. **Rocket Technology**: Initially invented during the Song Dynasty, gunpowder rockets were used in warfare and eventually laid the groundwork for the development of modern rocketry. Early rocket designs from China influenced military and scientific exploration in other cultures.

These contributions reflect China's legacy of innovation and exploration across diverse fields, with lasting impacts on global technology, culture, medicine, and commerce.

The story of the innovation of Paper production

The invention of paper is one of the most important innovations in human history, and it is traditionally credited to ancient China.

Early Writing Materials in China

Before paper, the ancient Chinese used various materials for writing, such as oracle bones (used during the Shang Dynasty, 1600–1046 BCE), bamboo strips, and silk. While these materials worked for record-keeping, they were cumbersome, heavy, and expensive. As a result, there was a need for a more efficient and cost-effective writing medium.

The Invention of Paper

The invention of paper is traditionally attributed to Cai Lun, an official of the Eastern Han Dynasty, around 105 CE. According to historical accounts, Cai Lun was the first to formalize the process of paper-making using plant fibers. He is credited with discovering a method of mixing tree bark, hemp, rags, and fishing nets with water to create a pulp, which was then pressed and dried to form sheets of paper.

The Process of Paper-Making

Cai Lun's process for making paper involved several key innovations:

1. **Pulping**: The raw materials (bark, hemp, rags, etc.) were broken down into fibers by soaking them in water and then pounding them to create a pulp.
2. **Molding**: The pulp was spread over a flat surface (usually a screen or mold) to form sheets.

3. **Pressing**: Excess water was removed by pressing the pulp, and the resulting sheets were left to dry in the sun.
4. **Finishing**: Once dried, the paper could be cut into the desired size and used for writing.

Cai Lun's method was a significant improvement over previous methods and allowed for the mass production of paper, which was lighter, more flexible, and more affordable than the materials that had been used previously.

Early Use and Spread of Paper

Initially, paper was used primarily for administrative and bureaucratic purposes. It allowed the Chinese to produce books, documents, and records more efficiently, which was important for the government and the military. Over time, paper became more widely used for writing, especially for calligraphy, painting, and printing.

The Spread of Paper to the West

The use of paper gradually spread from China to other parts of Asia, and eventually to the Middle East and Europe. The spread of paper was closely tied to the movement of people, especially via the Silk Road, the ancient trade route that connected East and West.

By the 8th century CE, paper-making had reached the Arab world, where it was further refined. The Arabs played a key role in spreading paper-making to Europe in the 12th century. By the 15th century, the invention of the printing press by Johannes Gutenberg in Germany, which used paper, revolutionized communication and knowledge-sharing across Europe.

***

The invention of paper had a profound impact on human history. It made the written word more accessible, facilitated the spread of knowledge, and helped to drive the development of printing, literature, education, and science. Paper became the primary medium for communication and record-keeping across the world, and it remains essential in modern society today.

In summary, while Cai Lun is often credited with the formal invention of paper around 105 CE, the process of making paper likely evolved over time, drawing on earlier experiments and innovations. The legacy of paper-making has endured for centuries and continues to shape the way we communicate and share knowledge today.

The story of the invention of Silk production

The story of silk production in China is a tale of mystery, innovation, and cultural significance that dates back over 5,000 years. The process of making silk was a closely guarded secret for centuries, and its invention is attributed to Empress Leizu, the wife of the Yellow Emperor (legendary Chinese ruler) during the Xia Dynasty (around 3000 BCE). According to Chinese legend, Empress Leizu discovered silk when a cocoon fell into her tea, and she observed how the delicate fibers unwound as she tried to remove it. She is often credited with inventing the technique of unwinding silk fibers and spinning them into threads.

The Beginning of Silk Production

The discovery of silk production marks one of the earliest examples of an ancient civilization mastering natural materials. The primary source of silk is the silkworm (Bombyx mori), whose larvae feed on mulberry leaves and spin protective cocoons made from a single long strand of silk. The Chinese developed techniques to harvest, unravel, and spin the fibers into silk threads. By carefully cultivating silkworms and mastering the process of reeling the silk from their cocoons, the Chinese established silk as an essential and highly prized material.

Secrecy and Early Trade

For centuries, the art of silk production was a closely guarded secret, and anyone who revealed the process was punished by death. This secrecy allowed China to dominate the global silk trade for over a thousand years. Silk became a symbol of luxury, wealth, and power within Chinese culture and beyond.

Silk and the Silk Road

By the time of the Han Dynasty (206 BCE–220 CE), silk production was well-established in China, and the fabric became a major export. The Silk Road, the trade route that connected China with Central Asia, the Middle East, and Europe, helped spread the knowledge of silk production, though the process remained secret for many centuries. The popularity of Chinese silk in distant lands led to its reputation as a luxurious fabric, sought after by rulers and elites across the world.

The Spread of Silk Production

Around the 6th century CE, the secret of silk production began to spread. According to legend, two Byzantine monks smuggled silkworm eggs and mulberry seeds out of China, eventually establishing silk production in the Byzantine Empire. Over time, the knowledge of silk-making spread to India, Japan, and eventually Europe, but the Chinese remained the leading producers for many centuries.

Today, China remains the world's largest producer of silk. The country's discovery and refinement of silk production played a crucial role in shaping its economy, culture, and influence across the ancient world. Silk continues to be a symbol of China's rich history and cultural heritage.

The story of the invention of Acupuncture

The story of acupuncture in China dates back thousands of years, with its roots deeply intertwined with the philosophy of Traditional Chinese Medicine (TCM). Acupuncture, which involves inserting fine needles into specific points on the body to stimulate energy flow and restore balance, is believed to have been developed as early as the Shang Dynasty (1600–1046 BCE), although its origins are surrounded by legend and gradual evolution over time.

Early Origins and Discovery

The earliest known references to acupuncture appear in ancient Chinese texts, particularly in the Yellow Emperor's Inner Canon (Huangdi Neijing), a foundational TCM text that dates back to around 300 BCE. The text outlines the principles of acupuncture and its role in balancing the body's energy, or qi (pronounced "chee"). According to TCM, qi flows through pathways in the body known as meridians, and the practice of acupuncture is meant to restore the proper flow of qi, thereby improving health and treating ailments.

While the exact origin story of acupuncture is unclear, some historical sources suggest that it may have developed from the ancient practice of using sharp stones or other materials to press on the body for therapeutic purposes. Over time, this evolved into the practice of inserting needles at specific points.

Acupuncture in the Warring States Period (475–221 BCE)

During the Warring States Period, acupuncture was further refined. It was in this period that acupuncture began to be more closely associated with the study of anatomy and the flow of energy in the body. The Yellow Emperor's Inner Canon played a key role in formalizing acupuncture theory, detailing how the body's energy system could be manipulated through specific acupuncture points.

Development and Spread

Acupuncture became a central part of Chinese medicine during the Han Dynasty (206 BCE–220 CE). The system of acupuncture points and meridians was further codified, and the practice became more widespread. At this time, acupuncture tools, such as metal needles, were refined, and the theory behind acupuncture was expanded upon.

During this period, acupuncture began to be integrated into medical practices alongside other therapies like herbal medicine and moxibustion (the burning of mugwort near the skin). It was also during the Han Dynasty that the five-element theory was introduced, linking acupuncture treatment to the balance of wood, fire, earth, metal, and water in the body.

Acupuncture's Evolution and Influence

Throughout the centuries, acupuncture continued to evolve, with further refinements in technique and understanding of the human body. By the Tang Dynasty (618–907 CE), acupuncture was considered a primary medical treatment in China. The practice spread to neighboring countries, including Korea, Japan, and Vietnam, where it was adapted and developed in unique ways.

Despite its enduring popularity in China, acupuncture faced periods of suppression, especially during the early 20th century, when Western medicine became more dominant. However, acupuncture experienced a resurgence in the mid-20th century, particularly after it was promoted internationally, especially following the 1970s when it gained attention in the West.

***

Today, acupuncture is practiced around the world and recognized as a therapeutic treatment for a variety of health conditions, including pain management, stress relief, and digestive issues. In 1979, the World Health Organization (WHO) officially recognized acupuncture as a viable treatment for certain conditions, solidifying its place in modern medicine.

Though acupuncture's exact origins are shrouded in the mists of time, it remains a central practice of Traditional Chinese Medicine and continues to be an influential and widely used healing method across the globe.

The story of the invention of the seismograph

The invention of the seismograph in China is attributed to the ancient Chinese polymath Zhang Heng (78–139 CE), a brilliant scientist, astronomer, mathematician, and inventor of the Han Dynasty. Zhang Heng's seismograph, known as the "earthquake weathercock" or "seismoscope," is considered the first device in the world capable of detecting and recording earthquakes.

The Need for Earthquake Detection

Ancient China, particularly during the Han Dynasty, was subject to frequent earthquakes, especially in the central and western regions. Despite the absence of modern technology, early Chinese scholars were keen to understand the causes of earthquakes and to find ways to detect them. At that time, earthquakes were often seen as ominous signs, and their detection was crucial for the safety of the population.

Zhang Heng's Seismoscope

In 132 CE, Zhang Heng designed and built his famous seismoscope, which could detect the tremors of an earthquake, even if the epicenter was far away. His seismoscope was a bronze vessel shaped like a large jar, with a pendulum inside that could move in response to seismic waves. The device was placed on a central stand and had a number of dragon-shaped mechanisms surrounding the vessel, each holding a ball in its mouth.

When an earthquake occurred, the seismic waves caused the pendulum inside the device to swing. This, in turn, triggered one of the dragons to release the ball into the mouth of a nearby frog figure positioned at the base of the seismoscope. The dragon's action indicated the direction of the earthquake's epicenter, and the fallen ball signaled the occurrence of the earthquake.

The seismoscope did not just detect the earthquake but also indicated its direction, which was a remarkable achievement, as it allowed the authorities to determine which part of the empire had been affected.

Zhang Heng's Achievement

Zhang Heng's seismoscope was groundbreaking for several reasons. First, it represented one of the earliest attempts to measure and record seismic activity. Unlike previous methods, which relied on human observation or simple, qualitative descriptions of earthquakes, Zhang Heng's seismoscope could detect and report earthquakes in a more systematic and objective manner.

According to historical records, Zhang Heng's seismoscope successfully detected an earthquake that occurred over 500 kilometers away from the device. This capability was considered a major technological breakthrough at the time and demonstrated Zhang Heng's deep understanding of physics and mechanics.

Influence

Although Zhang Heng's seismoscope was not widely used outside of China during his time, it laid the foundation for future developments in seismology. The seismoscope's concept of detecting and recording the movement of the Earth's surface through mechanical means influenced later innovations in seismology.

In the centuries following Zhang Heng's invention, seismographs evolved into more advanced devices, culminating in the modern electronic seismographs used today. However, Zhang Heng's seismoscope remains a significant milestone in the history of earthquake detection.

Zhang Heng's legacy as an inventor and scientist continues to be celebrated in China, where his seismoscope is remembered as one of the earliest and most impressive inventions in the history of science.

Zhang Heng's invention of the seismoscope in 132 CE marked a pivotal moment in the history of earthquake detection and geophysics. By creating a device that could detect and record earthquakes long before the advent of modern technology, Zhang Heng laid the groundwork for the science of seismology, and his work continues to inspire scientific progress today.

The story of the invention of mechanical clocks

The innovation of mechanical clocks in China is a remarkable journey that blends ancient Chinese ingenuity with the evolution of timekeeping technology. The development of these clocks spanned many centuries and involved the contributions of several key figures, including Yi Xing, Liang Lingzan, and Su Song. Their collective efforts laid the foundation for the development of precise mechanical timekeeping, which influenced both China and the broader world.

Early Timekeeping in China

Before the development of mechanical clocks, China had already invented sophisticated timekeeping methods, including the use of water clocks (clepsydra) and sundials. These early devices were primarily used to track time during the day or at night but had limitations in terms of accuracy. Despite their technological advancements, the need for more precise timekeeping devices became increasingly important, especially for astronomical and governmental purposes.

Yi Xing and Liang Lingzan (Tang Dynasty, 7th Century)

The story of the mechanical clock's innovation begins during the Tang Dynasty (618–907 CE), an era of great cultural and scientific achievements in China. Among the most significant figures in early mechanical clock development were Yi Xing, a Buddhist monk and astronomer, and his collaborator, Liang Lingzan, an engineer.

Yi Xing: The Theorist

Yi Xing is often recognized for his contributions to the astronomical and mathematical aspects of the mechanical clock. Yi Xing was a highly respected scholar and astronomer who understood the importance of precise timekeeping for the study of the stars and the management of governmental affairs. His key contribution was developing the astronomical principles that would later be incorporated into the design of the mechanical clock. Yi Xing worked with Liang Lingzan to combine these principles with mechanical engineering, leading to the creation of the first water-powered astronomical clock.

Liang Lingzan: The Engineer

While Yi Xing laid the groundwork with his scientific knowledge, Liang Lingzan played a crucial role in the physical design and construction of the clock. Liang was responsible for integrating mechanical gears and a water wheel system into the clock, allowing it to function continuously. The clock, which was powered by water wheels and used mechanical gears to regulate time, represented a breakthrough in timekeeping technology. It also included features like bell chimes to signal the passing of time, a significant leap forward from earlier timepieces.

Together, Yi Xing and Liang Lingzan created a clock that not only measured time but could also track astronomical events. This mechanical clock was one of the earliest examples of geared systems and is regarded as one of the first true mechanical clocks in history.

Su Song (Song Dynasty, 11th Century)

The most famous and advanced development in Chinese mechanical clockmaking occurred in the Song Dynasty (960–1279 CE) through the work of Su Song, a renowned polymath and government official. Su Song's water-driven astronomical clock tower, completed in 1088 CE, is considered one of the greatest achievements in the history of Chinese science and engineering.

Su Song's Clock Tower

Su Song's clock tower was an extraordinary feat of engineering. Located in the capital city of Kaifeng, the tower was a large, multi-tiered structure that featured a mechanical clock powered by a water wheel system. The clock not only kept time with great accuracy but also had a rotating celestial display that depicted the movements of the stars and planets.

Mechanical Gears and Escapement: Su Song's clock incorporated mechanical gears and an escapement mechanism, which regulated the movement of the clock's components, ensuring that the clock ran with precision. The escapement mechanism was a vital component for controlling the movement of the clock's gears, and it became a fundamental feature in later mechanical clocks, including those in Europe.

Water and Gear-driven System: The clock was powered by a water-driven gear system, which enabled it to operate continuously without human intervention. The clock's complex system of gears and water wheels was revolutionary for its time and showcased Su Song's understanding of hydraulics and mechanical engineering.

Astronomical Functions: In addition to telling the time, the clock also displayed astronomical phenomena, making it an important tool for scholars and astronomers of the time. The clock tower's celestial display could predict the position of the sun and stars, making it an advanced astronomical instrument.

Su Song's clock tower not only exemplified the technical prowess of Song Dynasty engineers but also influenced later developments in both China and the West.

The Legacy of Chinese Mechanical Clocks

The mechanical clocks developed by Yi Xing, Liang Lingzan, and Su Song had a profound impact on timekeeping technology both within China and beyond. While the mechanical clock would later be refined and developed further in Europe, the early Chinese innovations were among the first to incorporate gears, escapement mechanisms, and water-driven systems—technologies that would form the basis for Western clocks centuries later.

Su Song's clock tower, in particular, remained a model of technological advancement for centuries. The mechanical designs developed in China influenced the development of astronomical clocks in Persia and the Islamic world, and eventually made their way to Europe through cultural exchange along the Silk Road.

***

The innovation of mechanical clocks in China is a testament to the scientific and engineering prowess of ancient Chinese scholars. Yi Xing and Liang Lingzan pioneered the development of the first water-powered astronomical clocks, while Su Song's clock tower brought mechanical timekeeping to new heights. Their innovations in timekeeping influenced not only China but also the wider world, and their legacy continues to be felt in modern timekeeping technologies. These early inventions laid the groundwork for the precise clocks that would later shape the world's understanding of time.

Day 6: India

India has contributed significantly to science, mathematics, medicine, philosophy, and culture, with innovations that have shaped global civilization. Here are fifteen important innovations, explorations, and inventions introduced by Indians to the world:

1. **The Concept of Zero**: The Indian mathematician Brahmagupta formalized the concept of zero around the 7th century. This groundbreaking invention revolutionized mathematics and made modern arithmetic, algebra, and calculus possible.

2. **Decimal System**: Ancient Indian mathematicians developed the base-10 decimal system, which spread through the Middle East to Europe. This system is fundamental to modern mathematics and computing.

3. **Ayurveda**: As one of the oldest medical systems, Ayurveda includes treatments based on natural herbs, nutrition, and lifestyle. Its principles of wellness and balance influence alternative medicine practices globally.

4. **Yoga**: Originating over 5,000 years ago, yoga is a spiritual, physical, and mental discipline rooted in Indian philosophy. Today, it is practiced worldwide for its physical and mental health benefits.

5. **Sanskrit Grammar and Linguistics**: The ancient linguist Panini created a comprehensive grammar of Sanskrit in the 4th century BCE, considered the first descriptive grammar of any language. His work laid the foundation for linguistics and influenced modern languages.

6. **Algebra, Trigonometry, and Calculus**: Indian mathematicians, such as Aryabhata, Bhaskara II, and Madhava, made pioneering contributions in these fields. The Kerala school of mathematics developed calculus concepts centuries before Newton and Leibniz.

7. **Chess (Chaturanga)**: The game of chess originated in India around the 6th century as *Chaturanga*. It evolved as it spread to Persia, the Islamic world, and Europe, eventually becoming the modern game of chess.

8. **Plastic Surgery**: The ancient Indian physician Sushruta is often considered the father of plastic surgery. His work, *Sushruta Samhita*, describes surgical techniques, including reconstructive rhinoplasty, that are still in use today.

9. **Iron and Metallurgy**: Ancient Indian metallurgists developed advanced iron smelting techniques. The Iron Pillar of Delhi, which has not rusted for over 1,600 years, demonstrates this expertise and knowledge of metallurgy.

10. **Fibonacci Numbers**: Although associated with the Italian mathematician Fibonacci, the Fibonacci sequence was known to Indian mathematicians as part of Sanskrit poetry and Indian mathematical tradition well before it reached Europe.

11. **Binary Numbers**: The binary system, essential to modern computing, was first conceptualized by Indian mathematician Pingala in the 2nd century BCE. His work on prosody introduced the concept of binary numbers and influenced digital technology.

12. **Cotton and Textile Production**: India was one of the earliest civilizations to cultivate cotton and develop weaving techniques, producing fine cotton textiles that were traded along ancient trade routes. India's textile innovations influenced fashion and textiles worldwide.

13. **Ink**: Ancient Indians developed early forms of ink, used for writing on scrolls and manuscripts, as far back as the 4th century BCE. This black ink, known as *masi*, was used in early literature and influenced writing materials.

14. **Diamonds and Gemology**: India was the first country to mine diamonds, with historical records dating back over 2,500 years. Indian expertise in diamond cutting and gemology laid the groundwork for the global diamond industry.

15. **Board Games (Snakes and Ladders)**: Originally known as *Moksha Patamu*, this game was created in ancient India to teach morality and the journey of life. It was later adapted in the West as "Snakes and Ladders" and remains popular worldwide.

These contributions from India highlight a rich legacy of knowledge, innovation, and cultural influence that has had a profound impact on global civilization.

The story of the invention of Zero

The invention of zero is one of the most remarkable contributions of ancient Indian mathematicians to the world, fundamentally transforming mathematics, science, and philosophy. Although the concept of zero seems simple and ubiquitous today, its development was a groundbreaking achievement that enabled advances in various fields and laid the foundation for modern number systems and computing.

Early Beginnings: The Concept of Nothingness

The notion of "nothingness" or "emptiness" was not unique to India. Many ancient cultures, including the Babylonians and the Greeks, had some representation of "nothing" or placeholders in their counting systems, often to differentiate between values (e.g., the difference between 10 and 100). However, these cultures did not fully develop the concept of zero as a number with mathematical properties.

In ancient Indian thought, particularly in Hindu and Buddhist philosophies, the idea of shunya (Sanskrit for "void" or "emptiness") played a significant role. This philosophical concept of emptiness may have inspired the development of zero, as Indian thinkers began to explore the mathematical implications of "nothingness" in practical and abstract ways.

The Development of Zero as a Symbol and a Number

The first recorded use of zero as a symbol and a mathematical concept is found in India. The use of zero in Indian mathematics evolved over centuries and is most closely associated with the work of mathematicians in the Gupta Empire period (circa 5th–6th centuries CE).

1. **Ancient Texts and Inscriptions**: The earliest known inscription of the zero symbol appears in the Bakhshali Manuscript, an ancient Indian mathematical text that dates back to somewhere between the 3rd and 7th centuries CE. This manuscript used a small dot to represent zero as a placeholder, marking one of the earliest uses of zero in positional notation.

2. **Brahmagupta's Breakthrough (7th century)**: The Indian mathematician Brahmagupta is often credited with establishing

zero as a true number with its own set of rules. In 628 CE, Brahmagupta wrote the Brahmasphutasiddhanta, an astronomical text where he not only used zero but also described how to operate with it. He defined zero's properties, such as:

- **Addition and subtraction** with zero: $x+0=x$ and $x-0=x$.
- **Multiplication**: Any number multiplied by zero is zero.
- However, division by zero remained a concept that was not fully resolved, as it leads to undefined results.

Brahmagupta's rules for zero and other mathematical innovations allowed Indian mathematicians to work with complex equations and calculations previously impossible in ancient mathematics.

The Spread of Zero to the Islamic World and Europe

The concept of zero, along with the Indian numeral system (0–9), began to spread beyond India through cultural and trade exchanges.

1. **Transmission to the Islamic World**: Indian mathematics reached the Islamic world through scholars and traders. The Persian mathematician Khwarizmi (circa 780–850 CE), often called the "father of algebra," learned about Indian mathematics and zero and incorporated it into his own work. His book, which introduced the Indian numeral system and zero, greatly influenced Islamic mathematics.

2. **Arrival in Europe**: From the Islamic world, the concept of zero spread to Europe, particularly after the Latin translations of Al-Khwarizmi's works. The Italian mathematician Fibonacci (Leonardo of Pisa) introduced zero and the Hindu-Arabic numeral system to Europe in the 13th century through his book Liber Abaci (1202), which popularized these ideas among European scholars.

3. **Resistance and Acceptance**: Initially, zero faced resistance in Europe due to its association with the unfamiliar Indian-Arabic numeral system and its implications for religious and philosophical beliefs. However, as the utility of zero in mathematics and commerce became evident, it gradually gained acceptance and

became an essential component of European and Western mathematics.

Importance of Zero in Mathematics and Beyond

The invention of zero revolutionized mathematics, science, and technology, providing a foundation for developments that would shape the modern world.

1. **Place-Value System**: Zero enabled the place-value system in mathematics, where the position of each digit in a number indicates its value (e.g., ones, tens, hundreds). This system is foundational to all arithmetic and calculations in base-10 notation.

2. **Algebra and Calculus**: Zero allowed the development of algebra by providing a way to solve equations, and it paved the way for calculus and other branches of higher mathematics. Algebraic and calculus-based models are used extensively in physics, engineering, and economics.

3. **Negative Numbers and Infinity**: With zero, mathematicians could now explore negative numbers and concepts of infinity and limits, expanding mathematical analysis. These ideas were crucial for the development of theories in modern physics, including Einstein's theories of relativity.

4. **Binary System and Computers**: The binary system, which is the foundation of all digital computers, relies on only two numbers—zero and one. This digital binary system enables all modern computing, from simple calculators to complex supercomputers.

5. **Philosophical and Scientific Impact**: Zero also influenced philosophical and scientific thinking, challenging human conceptions of existence, emptiness, and infinity. It contributed to discussions about the nature of the universe, the structure of matter, and concepts of voids in physics.

***

The Indian invention of zero was a revolutionary achievement that reshaped the mathematical landscape and laid the foundation for scientific and technological progress. It is considered one of the greatest innovations in human history, impacting mathematics, philosophy, computer science, and beyond. Zero transformed human knowledge by enabling complex calculations, inspiring new fields of study, and facilitating the development of tools we rely on today.

The Indian numeral system, with zero as its cornerstone, is now used universally, underscoring the enduring legacy of ancient Indian mathematicians and the profound influence of their ideas on the modern world.

The story of the invention of Yoga

The story of the invention of Yoga is deeply rooted in ancient Indian philosophy and spirituality, with origins dating back thousands of years. Yoga is not an invention in the conventional sense, but rather a profound practice that evolved over time, encompassing physical, mental, and spiritual elements to help individuals achieve harmony, balance, and enlightenment.

Ancient Beginnings of Yoga (Pre-Vedic Period)

The origins of Yoga can be traced back to the Indus Valley Civilization (around 3000 BCE), where archaeological evidence suggests early forms of meditative practices. Ancient seals and artifacts from the Indus Valley, such as a famous seal depicting a figure seated in a meditative posture, indicate that people in the region were practicing techniques resembling meditation and yoga as early as 5,000 years ago.

However, Yoga as we know it today began to take shape in the Vedic period (1500–500 BCE), when the Vedas, the oldest sacred texts of Hinduism, were written. These texts contain hymns and rituals meant to invoke spiritual growth and the connection between humans and the divine.

Yoga in the Upanishads (800-400 BCE)

As Indian spirituality and philosophy evolved, Yoga became more refined in the Upanishads, a collection of mystical and philosophical texts that explore the nature of existence, the soul, and the universe. These texts discuss concepts such as meditation, self-realization, and union with the divine, which are central to the practice of Yoga.

During this period, meditation (dhyana) and ascetic practices were key components of spiritual development. These early practices focused on attaining spiritual enlightenment and inner peace, and the term Yoga began to emerge more explicitly to describe the union between the individual self (Atman) and the universal consciousness (Brahman).

The Rise of Classical Yoga (200 BCE–500 CE)

The formalization of Yoga continued during the Classical period, and one of the most influential figures in this development was the sage Patanjali. Around 200 BCE, Patanjali wrote the Yoga Sutras, a compilation of aphorisms (short verses) that systematically outlined the philosophy and practice of Yoga. The Yoga Sutras are considered one of the most important texts in the history of Yoga and form the foundation of classical Yoga.

Patanjali's system of Yoga is known as Raja Yoga, or the "royal path," and it emphasizes the mind and meditative practices. The Eight Limbs of Yoga (Ashtanga Yoga) outlined in the Yoga Sutras include practices such as ethical conduct (Yama and Niyama), physical postures (Asana), breathing techniques (Pranayama), and meditation (Dhyana). These teachings focused on controlling the mind and senses, ultimately leading to self-realization and spiritual liberation (Moksha).

The Bhagavad Gita and the Development of Bhakti Yoga (500 BCE–300 CE)

The Bhagavad Gita, written around the 2nd century BCE, is another key text in the development of Yoga. It presents a conversation between the prince Arjuna and the god Krishna, in which Krishna teaches Arjuna the paths of Karma Yoga (the Yoga of selfless action), Bhakti Yoga (the Yoga of devotion), and Jnana Yoga (the Yoga of knowledge). The Bhagavad Gita emphasizes that Yoga is not just a physical discipline but a comprehensive spiritual practice that helps one to unite with the divine.

The concept of Bhakti Yoga (the path of devotion) became a central practice in later forms of Yoga, highlighting the importance of love and devotion to God as a way to achieve spiritual liberation.

The Rise of Hatha Yoga (9th–15th Century)

In the medieval period, the practice of Hatha Yoga emerged as a system of physical exercises designed to purify and strengthen the body in preparation for higher spiritual practices. Hatha Yoga emphasized the practice of physical postures (Asanas) and breathing techniques (Pranayama) to achieve physical health and control over the body.

One of the most important texts on Hatha Yoga is the Hatha Yoga Pradipika, written in the 15th century by the sage Svatmarama. The Hatha Yoga Pradipika is a guide to the practice of asanas, pranayama, mudras (hand gestures), and bandhas (body locks), all aimed at controlling the body and mind in pursuit of spiritual awakening.

Hatha Yoga served as a foundation for modern forms of physical yoga practiced today, focusing on improving physical health, flexibility, and mental clarity.

Yoga in the Modern World (19th–20th Century)

Yoga underwent significant changes in the 19th and 20th centuries, as it began to spread beyond India and gain global recognition. The physical aspects of Yoga became especially popular in the West, with the rise of Hatha Yoga and other physical forms of yoga practiced for fitness and health. Key figures in this transition included Swami Vivekananda, who introduced Yoga to the West at the 1893 World's Parliament of Religions in Chicago, and Tirumalai Krishnamacharya, who is often credited with modernizing and systematizing yoga postures (asanas).

Other influential figures like B.K.S. Iyengar, Pattabhi Jois, and Indra Devi helped spread Yoga worldwide in the 20th century. As Yoga became more popular, it was embraced not only as a spiritual practice but also as a physical discipline that promotes wellness and stress relief.

The invention of Yoga cannot be attributed to a single individual or moment, as it evolved over millennia in India. It emerged as a spiritual practice aimed at achieving self-realization and union with the divine, and it has been shaped by numerous philosophies, texts, and teachers throughout history. From its early roots in the Vedic and Upanishadic traditions to its codification by Patanjali in the Yoga Sutras, Yoga has continually evolved to encompass both physical and spiritual practices.

In the modern era, Yoga has become a global phenomenon, practiced by millions of people worldwide for its physical, mental, and spiritual benefits. Its deep philosophical and spiritual roots in India continue to guide the practice, making it a profound and transformative discipline for those seeking inner peace, balance, and enlightenment.

The story of the invention of Chess

The story of the invention of chess is rooted in ancient India, where it evolved over centuries into the game we recognize today. The origins of chess are linked to a game known as chaturanga, which was played in India around the 6th century CE.

Chaturanga: The Ancient Predecessor

The earliest form of chess was chaturanga, a strategic board game played on an 8x8 grid similar to the modern chessboard. The name "chaturanga" comes from the Sanskrit word "chatur" (meaning four) and "anga" (meaning parts or limbs). This refers to the four divisions of the ancient Indian army: infantry, cavalry, elephants, and chariots. The game represented these divisions with different pieces:

- **Infantry**: represented by pawns
- **Cavalry**: represented by knights
- **Elephants**: represented by bishops
- **Chariots**: represented by rooks

Chaturanga was a game of military strategy, and it is believed to have been played by Indian royalty and military leaders to practice tactics and warfare strategies. The game was not just a pastime but also served as a tool for educating and training warriors and tacticians.

Evolution and Spread of Chaturanga

Chaturanga gained widespread popularity in India, and by the 7th century, it had spread to Persia, where it became known as shatranj. The game continued to evolve as it made its way westward into the Islamic world and, eventually, to Europe.

In Persia, the pieces of chaturanga underwent some modifications, and the rules were adjusted. For example, the "elephant" (which moved two squares diagonally) was replaced with the bishop, and the queen in shatranj had more limited movement compared to its modern counterpart.

The Transition to Modern Chess

By the time the game reached Europe, particularly in Spain and Italy, it underwent further modifications. The queen gained its modern ability to move any number of squares in any direction, and the bishop was given its current diagonal movement. These changes made the game faster and more dynamic, moving closer to the form of chess we know today.

In the 15th century, the game began to be referred to as "chess", derived from the Persian word "shah", meaning "king," which was shouted when the king was threatened or "checkmated." The rules were further standardized during this period, and chess spread throughout Europe, where it gained widespread popularity among nobility.

Chess in India Today

Although India is not as associated with the game in the modern era, it remains proud of its ancient connection to the game. India has produced some of the world's greatest chess players, such as Viswanathan Anand, a five-time world chess champion, and Baba Raman and K. K. Venkatraman, who have further reinforced India's historical role in the world of chess.

The invention of chess in India, through the ancient game of chaturanga, has had a profound impact on global culture. As it evolved over centuries and spread across the world, it became a game of intellect, strategy, and skill that continues to captivate people of all ages. Chess is not only a game but also a representation of military strategy, intellectual pursuit, and cultural exchange, with its roots firmly planted in ancient India.

The story of the invention of Binary system

The story of the binary system's invention is credited to the Indian mathematician Pingala, who lived around the 2nd century BCE. While binary systems are commonly associated with modern computers and digital technology, Pingala's work laid the foundation for these concepts long before their practical use in the modern world.

Pingala and the Early Development of the Binary System

Pingala is best known for his work in combinatorics and mathematics, particularly in the field of Sanskrit prosody (the study of poetic meters). His text, the Chandas Shastra (also known as the "Science of Verses"), is one of the earliest known works to explore the mathematical aspects of rhythm and meter in poetry.

In his Chandas Shastra, Pingala introduced a system of enumerating and arranging syllables, which he used to describe the patterns in Sanskrit verse. His approach involved breaking down rhythmic patterns into two possible elements: a short syllable (referred to as "laghu") and a long syllable ("guru").

The Binary System and its Connection to Sanskrit Meter

While analyzing these syllabic patterns, Pingala used a binary-like notation to represent the rhythmic units. In his system, a short syllable was represented by "1" (a binary "on" state), and a long syllable was represented by "0" (a binary "off" state). This system of two distinct states—short and long syllables—had similarities to the modern binary system used in digital computing, where information is represented using two states, 0 and 1.

Pingala's binary approach was essentially a way to simplify the enumeration of possible combinations of syllabic patterns in poetry. The binary-like representation helped him organize and identify the different combinations of short and long syllables to form metrical patterns in Sanskrit poetry.

The Concept of Binary Numbers

Pingala's binary system can be seen as an early precursor to the modern binary number system used in computer science. In the Chandas Shastra, he introduced a form of combinatorics where sequences of 1s and 0s were used to represent different possible combinations. This system of counting and arrangement in binary patterns is remarkably similar to the principles used in digital encoding in computers, where data is stored and processed in binary form.

While Pingala's binary system was not directly connected to the modern concept of digital computing, his use of binary-like sequences to represent syllables and rhythms was an important step in the development of mathematical logic and systems that later contributed to the creation of binary number systems.

Influence on Later Developments

Although Pingala's binary system was primarily focused on prosody, it laid the groundwork for later developments in mathematics, especially in the field of combinatorics and numerical systems. The concepts he introduced would influence later mathematicians and philosophers, including Rene Descartes, Gottfried Wilhelm Leibniz, and others, who would formally develop the modern binary system and apply it to the field of mathematics and computing.

In the 17th century, the German philosopher Leibniz independently developed a modern binary system based on the work of Pingala and others. Leibniz's version of binary notation became central to the development of digital computers, as it forms the basis of how computers represent and process data today.

The binary system that Pingala conceptualized in the 2nd century BCE was a groundbreaking development in mathematical thought. His work, originally designed for the study of Sanskrit meter and poetry, laid the foundation for a two-state system that later played a key role in the development of digital computing. Though Pingala's binary system was not initially recognized for its computing implications, it represents a fascinating intersection of mathematics, linguistics, and computer science, and his contribution is still acknowledged as a precursor to the binary system used in modern technology.

Day 7: Pakistan

Pakistan has contributed several significant innovations and advancements, particularly in science, technology, and social initiatives, that have made an impact both locally and internationally. Here are ten important contributions from Pakistanis:

1. **Development of the Ommaya Reservoir**: Neurosurgeon Ayub Ommaya invented the Ommaya Reservoir, a device for delivering drugs directly to the brain. This innovation has had a lasting impact on neurosurgery and the treatment of brain tumors and other central nervous system diseases.

2. **Electroweak Unification Theory (Contributions of Dr. Abdus Salam)**: Pakistani physicist Dr. Abdus Salam, a Nobel laureate, was one of the key contributors to the electroweak unification theory, which explained the fundamental forces governing subatomic particles. His work advanced the field of particle physics and inspired researchers globally.

3. **National Database and Registration Authority (NADRA)**: Pakistan's NADRA developed one of the most advanced biometric citizen identification systems in the world. This system has served as a model for biometric and identity management systems worldwide, improving security and access to government services.

4. **Integrated Satellite Technology (SUPARCO)**: Pakistan's space agency, SUPARCO, has made significant contributions to satellite technology in the region, launching satellites for communication, weather forecasting, and research. This has advanced space capabilities and satellite usage in South Asia.

5. **The First Computer Virus**: In 1986, two Pakistani brothers, Basit and Amjad Farooq Alvi, created the first known computer virus, dubbed the "Brain" virus. Although it was initially intended to protect their software from piracy, it led to the birth of the field of cybersecurity and antivirus software development.

These innovations showcase Pakistan's contributions in social services, healthcare, space technology, and scientific research, with lasting impacts on both national and global scales.

The story of the invention of the Computer Virus

The story of the first computer virus created by two Pakistani brothers, Amjad and Basit Farooq Alvi, is a fascinating tale from the early days of personal computing. In 1986, these brothers inadvertently became the creators of what is now considered the first widely known computer virus in history, known as the Brain virus. This virus marked the beginning of cybersecurity as a field and introduced the world to the new risks associated with computer technology.

Background: The Beginnings of Personal Computing in Pakistan

Amjad and Basit Farooq Alvi were skilled programmers and computer enthusiasts in Lahore, Pakistan. In the mid-1980s, personal computers were starting to become more popular around the world, but in Pakistan, they were still quite rare. The Alvi brothers ran a small computer repair and software business, where they provided services for local companies and individuals.

However, the Alvi brothers faced a problem common to many early software developers: piracy. Software piracy was rampant, with people copying programs and distributing them without permission. The brothers saw their software, developed after hours of work, being copied and used without any credit or compensation. They wanted to find a way to discourage piracy and track the unauthorized use of their software, hoping to raise awareness and protect their intellectual property.

Creation of the Brain Virus

To tackle the piracy issue, the Alvi brothers came up with a unique solution—a piece of software that could "infect" a computer's boot sector, which is the part of a disk that tells a computer how to start up. Their goal wasn't to damage users' data or cause harm but to let unauthorized users know that they were using pirated software.

The Brain virus was designed to copy itself to the floppy disk (the main storage medium at the time) of the infected computer and, from there, spread to other computers when users shared disks. Unlike the destructive malware that would emerge later, Brain did not delete files or corrupt data. Instead, it embedded a message in the infected disk's boot sector:

Welcome to the Dungeon/(c) 1986 Basit & Amjad/VIRUS_SHOE RECORD v9.0/ Dedicated to the dynamic memories of millions of virus owners!!

The message also contained the Alvi brothers' contact information, including their names, address, and phone numbers, so affected users could reach out to them directly. This message clearly indicated that the brothers hadn't intended to remain anonymous—they hoped users would contact them to understand why their software was being copied illegally.

How the Brain Virus Worked

The Brain virus worked by overwriting the boot sector of floppy disks with a modified version of itself. Every time an infected floppy disk was inserted into a computer, the virus would activate and load itself into the computer's memory. From there, it would wait for the next floppy disk to be inserted, and it would copy itself onto that disk as well. This meant that any infected disk, when used in another computer, would carry the virus and potentially spread it to other computers.

Although the virus itself did not harm data, it slowed down the infected disks, making them slightly more difficult to use, which prompted users to inspect their disks. This eventually led to the realization that something unusual was happening.

Unexpected Global Spread and Impact

The Alvi brothers did not anticipate that the Brain virus would spread beyond their local community, let alone become an international phenomenon. However, since people frequently exchanged floppy disks, the virus began spreading globally, reaching countries as far away as the United States and Europe.

The global spread of the Brain virus shocked the brothers, who received calls from people around the world. Some were curious, others were angry, and some simply wanted to understand what had happened to their computers. Many in the computing world were also fascinated and alarmed—this was a new type of cyber threat that challenged existing ideas about computer security.

The Legacy of the Brain Virus

The Brain virus is often regarded as the first computer virus to gain worldwide attention. While it was created with a specific goal (to deter piracy and encourage users to buy original software), it unintentionally highlighted the vulnerabilities in computer systems and demonstrated how easily malicious software could spread across networks.

Its legacy includes several important impacts:

1. **Cybersecurity Awareness**: The Brain virus helped raise awareness about computer security and the potential for malicious software to spread. This virus marked the beginning of the cybersecurity field, as developers and users realized they needed to protect their systems from similar threats.

2. **The Rise of Antivirus Software**: The Brain virus and other early viruses led to the creation of the first antivirus programs. Companies specializing in cybersecurity, such as McAfee and Symantec, were founded to create software that could detect and remove viruses.

3. **Ethics of Hacking and Software Piracy**: The Brain virus sparked discussions about the ethics of hacking and the importance of protecting intellectual property. While the Alvi brothers had good intentions, the unintended spread of Brain showed that even well-meaning code could have unforeseen consequences.

4. **Awareness of Global Connectivity**: Brain demonstrated how quickly information—and in this case, malicious code—could travel across borders. It was one of the earliest instances showing how interconnected and vulnerable computer networks were, even before the widespread use of the internet.

The Alvi Brothers Today

In later interviews, the Alvi brothers expressed surprise at the way the Brain virus spread and the legacy it left behind. While they had only intended to protect their software, they inadvertently started a new chapter in the history of computers. They continued their work in the IT industry, running a successful internet service provider business in Pakistan.

***

The Brain virus was not designed to cause harm, yet it became a wake-up call that revealed the risks of digital connectivity. Created by two young Pakistani brothers, this virus unintentionally inaugurated the era of computer viruses and cybersecurity. Today, as we navigate a world filled with sophisticated malware and cyber threats, the Brain virus serves as a reminder of the humble beginnings of digital security and the profound influence of curiosity and experimentation in shaping technology's future.

The story of the invention of the Ommaya Reservoir

The Ommaya Reservoir is a groundbreaking medical device invented by Dr. Ayub Ommaya, a pioneering Pakistani neurosurgeon and researcher. This device revolutionized the treatment of brain tumors and other conditions requiring direct drug delivery to the brain, significantly improving patient care and survival rates for those with brain-related diseases.

Dr. Ayub Ommaya: Early Life and Background

Dr. Ayub Ommaya was born in Rawalpindi, Pakistan, in 1930. From an early age, he demonstrated a passion for science and medicine, which eventually led him to pursue medical studies. Dr. Ommaya attended King Edward Medical College in Lahore, where he excelled and graduated with honors. Driven by a thirst for knowledge and a desire to make meaningful contributions to medicine, he later moved to the United Kingdom to complete his advanced training and then to the United States to further his career in neurosurgery.

In the U.S., Dr. Ommaya became a respected figure in the field of neurosurgery. He worked as a senior researcher at the National Institutes of Health (NIH) and became known for his innovative approach to treating traumatic brain injuries and brain cancer. It was in this context that he developed the Ommaya Reservoir—a device that would mark a major breakthrough in neurosurgery and cancer treatment.

The Problem: Treating Brain Tumors and Other Neurological Conditions

During the 1960s, treating brain tumors was exceptionally challenging. The blood-brain barrier, a protective barrier that shields the brain from harmful substances in the bloodstream, also made it difficult to deliver chemotherapy drugs directly to the brain. While intravenous chemotherapy was effective in other parts of the body, it often had little impact on brain tumors because most drugs could not cross the blood-brain barrier.

Doctors and researchers sought ways to bypass this barrier and deliver drugs directly to the brain. Traditional methods, such as surgery or repeated injections into the spinal cord, were highly invasive and carried significant risks. Dr. Ommaya saw the need for a safer, more efficient method to administer drugs directly into the cerebrospinal fluid (CSF), which could then circulate the medication throughout the brain.

The Invention of the Ommaya Reservoir

To address this challenge, Dr. Ommaya designed a device that could deliver medication directly to the brain in a controlled, minimally invasive way. This device, known as the Ommaya Reservoir, consists of a small dome-shaped silicone reservoir implanted beneath the scalp. It is connected to a thin catheter that is inserted into the brain's ventricular system, where cerebrospinal fluid flows.

The reservoir serves as an access point for doctors to deliver medication directly to the brain. By using a needle to inject drugs through the skin and into the reservoir, medical professionals could release drugs into the CSF, which would distribute the medication throughout the brain. This system allowed for controlled, repeatable, and targeted drug delivery without the need for multiple surgeries or invasive procedures.

How the Ommaya Reservoir Works

The Ommaya Reservoir operates as a simple yet elegant solution:

1. **Implantation**: The reservoir is surgically implanted beneath the patient's scalp, and the catheter is positioned in the brain's ventricles.

2. **Injection**: To administer medication, doctors use a needle to inject the drug through the scalp and into the reservoir.
3. **Drug Distribution**: From the reservoir, the drug flows through the catheter and is released into the cerebrospinal fluid, allowing it to circulate throughout the brain and reach areas affected by cancer or other conditions.

Applications and Impact of the Ommaya Reservoir

The Ommaya Reservoir had far-reaching implications for brain cancer treatment and neurological care. By enabling a safer and more effective means of delivering chemotherapy directly to the brain, it significantly improved treatment outcomes for patients with brain tumors and metastases (cancers that spread to the brain from other parts of the body).

Some of the key applications and benefits include:

- **Treatment of Brain Tumors**: The reservoir allowed for direct chemotherapy treatment of brain tumors, increasing the effectiveness of the drugs while reducing the side effects associated with intravenous chemotherapy.

- **Management of Brain Infections**: The Ommaya Reservoir proved useful for treating severe brain infections by delivering antibiotics directly to the affected areas.

- **Cerebrospinal Fluid Sampling**: The reservoir enabled doctors to easily obtain cerebrospinal fluid samples, which helped in diagnosing and monitoring conditions affecting the brain and nervous system.

This device not only extended the lives of many cancer patients but also improved their quality of life by reducing the discomfort and risks associated with more invasive procedures. Dr. Ommaya's invention demonstrated how a relatively simple device could provide profound benefits for patients and transformed the standard of care in neuro-oncology.

***

Dr. Ayub Ommaya's invention is considered a landmark in neurosurgery and oncology, and the Ommaya Reservoir continues to be used worldwide. It laid the groundwork for further advances in targeted drug delivery, contributing to the development of newer devices and methods that continue to improve patient care.

Dr. Ommaya also made significant contributions to the study of traumatic brain injuries, particularly in understanding the effects of brain injuries in car accidents. His research in these areas contributed to improvements in road safety and brain injury prevention.

Day 8: Iran

Iranian civilization, with its rich history, has significantly contributed to global advancements in science, medicine, engineering, arts, and philosophy. Here are fifteen of the most important innovations, explorations, and inventions introduced by Iranians:

1. **Algebra**: The Persian mathematician Khwarizmi, often regarded as the "father of algebra," wrote *Kitab al-Jabr*, from which we derive the term "algebra." His work laid the foundation for modern mathematics.

2. **Algorithm**: Also credited to Khwarizmi, the concept of the "algorithm" (a step-by-step method for calculations) originated from his works, greatly influencing computer science and mathematics.

3. **Windmills**: Ancient Iranians invented the earliest windmills over a thousand years ago. These vertical-axis windmills were used to grind grain and pump water, especially in the arid regions of Persia.

4. **The Persian Garden**: The concept of the Persian garden, with its unique irrigation techniques and aesthetic structure, inspired garden designs across the Middle East, India, and even Europe, notably influencing the design of the Taj Mahal gardens.

5. **Surgical Techniques and Anesthesia**: Avicenna (Ibn Sina), a Persian polymath, advanced surgical methods and was among the first to introduce anesthesia techniques. His *Canon of Medicine* was used in Europe and the Middle East for centuries.

6. **Battery (The Baghdad Battery)**: Discovered near Baghdad, this ancient artifact is believed to be an early form of a battery, possibly used in medical or electroplating processes, showcasing early Persian innovations in electrical concepts.

7. **Qanat System**: This ancient irrigation technique, involving a series of well-like vertical shafts connected by gently sloping tunnels, allowed groundwater to be transported over long distances and enabled agriculture in arid areas. It spread across the Middle East and North Africa.

8. **Distillation and Alcohol Extraction**: Persian alchemists pioneered early methods of distillation, which laid the foundation for chemistry. They also created and refined the process of alcohol extraction for medicinal purposes.

9. **Astronomical Instruments**: Persian astronomers invented tools such as the astrolabe and contributed to observational astronomy. Observatories like the one in Maragha influenced Islamic and European astronomers alike.

10. **Ethics and Logic in Medicine**: Avicenna introduced ethical guidelines and logical methods for clinical medicine. His *Canon of Medicine* advocated for evidence-based practice and systematic medical diagnoses, influencing modern medical ethics.

11. **Early Forms of Antibiotics**: Rhazes (Razi), a Persian polymath, identified infections and developed early antiseptics. His innovative approaches to medicine included understanding that disease could be contagious and the benefits of antiseptics.

12. **Sulfuric Acid Production**: Persian alchemists developed methods to distill and purify sulfuric acid, later becoming crucial in chemistry and industrial processes, such as the refining of gold and silver.

13. **Polo (Chogan)**: Originating in ancient Persia, polo was not only a sport but also a means of military training for cavalry. The game spread to China, India, and later the British Empire.

14. **The Persian Carpet**: Known for their intricate designs and symbolism, Persian carpets became globally admired works of art and craftsmanship, influencing textile industries and artisans worldwide.

15. **The Postal System (Chapar Khaneh)**: The Achaemenid Empire developed one of the earliest postal systems to maintain communication across the empire. This efficient relay system inspired similar models used in the Roman Empire and other civilizations.

16. **Gondishapur University**: Gondishapur University, founded in 361 CE in ancient Iran, is considered one of the world's first universities. Renowned for its contributions to medicine,

philosophy, and science, it attracted scholars from across the ancient world, blending Greek, Persian, and Indian knowledge.

These contributions showcase Iran's role as a historical and intellectual hub, whose influence on science, culture, and technology has been widespread and enduring.

The story of the invention of Sulfuric acid production

The invention of sulfuric acid production is a notable achievement attributed to early Iranian alchemists. Known as one of the most important chemicals in industrial history, sulfuric acid is a powerful substance with countless applications in fields from chemistry to manufacturing. The journey of its discovery and refinement began in ancient Persia, where scientists and scholars delved into the natural properties of various substances in their quest to understand the world.

Background: Early Alchemy in Persia

The Persian Empire was a center of scientific knowledge and innovation, especially in fields like alchemy, medicine, and metallurgy. Persian scholars during the Islamic Golden Age (8th–14th centuries) were highly curious about the nature of materials and the transformation of substances. They laid the groundwork for modern chemistry, and their work spread throughout the Islamic world, ultimately influencing Europe and the development of modern scientific methods.

One of the earliest known Persian scholars involved in the study of sulfuric acid was Jābir ibn Hayyān (known as Geber in the West). Jābir, often called the "father of chemistry," wrote extensively on chemical processes and pioneered experimental methods in his writings. His work established a foundation for the production of several acids, including sulfuric acid.

The Process: Early Methods of Producing Sulfuric Acid

Sulfuric acid is created through chemical reactions involving sulfur and oxygen in the presence of a catalyst. Ancient Persian scholars developed a method for producing this acid through a process involving heating materials like green vitriol (iron sulfate) or alum in glass or clay containers. When heated, these substances released sulfur dioxide gas, which, when combined with water, produced sulfuric acid.

The two primary processes for early sulfuric acid production were:

1. **Heating Sulfur with Nitre (Potassium Nitrate)**: By burning sulfur in the presence of potassium nitrate, early chemists could produce a vapor that condensed into sulfuric acid. This method required skill to control the heat and ensure the sulfuric vapor was collected effectively.

2. **Distillation of Vitriols (Sulfates)**: The Persians also developed the technique of distilling "vitriol" minerals, such as iron sulfate (green vitriol) and copper sulfate (blue vitriol). When heated, these minerals decomposed and released sulfuric acid vapors that could be dissolved in water. This process became the basis of sulfuric acid production in Persia and was documented in various alchemical manuscripts.

Jābir ibn Hayyān's Contributions

Jābir ibn Hayyān is often credited with the refinement of sulfuric acid production techniques. His writings introduced chemical concepts and outlined laboratory processes that allowed for more systematic experimentation. He created recipes for sulfuric acid, along with hydrochloric and nitric acids, which became known as the "spirits" of alchemy.

Jābir's method involved using specific ratios of ingredients and carefully controlling the heating process to extract sulfuric acid in a purer form. These procedures, recorded in his extensive works, spread across the Islamic world and later influenced European alchemy. His methods made it possible to obtain concentrated sulfuric acid, which was crucial for several applications, including the creation of aqua regia, a mixture capable of dissolving gold.

Importance of Sulfuric Acid in Ancient Times

Sulfuric acid became a valuable substance in ancient alchemy for several reasons:

1. **Dissolving Metals**: Sulfuric acid's ability to dissolve metals and other substances made it a powerful tool in the study of metals and alloys, essential for advancing metallurgical practices.

2. **Catalyst in Alchemical Reactions**: It enabled the creation of new chemical compounds, facilitating experiments that led to the development of early medicines, pigments, and other useful materials.

3. **Foundation for Future Acid Production**: Jābir's methods for creating sulfuric acid laid the groundwork for producing other acids, such as nitric acid, which further expanded the capabilities of alchemists and chemists.

4. **Development of Laboratory Equipment and Techniques**: Producing sulfuric acid required controlled heating and distillation techniques, leading to the development of glassware and other apparatus used in early chemistry labs.

***

The methods for producing sulfuric acid pioneered by Persian alchemists were instrumental in the spread of alchemical knowledge. Through trade and scholarly exchange, knowledge of sulfuric acid and other chemical techniques reached Europe. By the time of the Renaissance, European alchemists had access to sulfuric acid, often calling it "oil of vitriol," and it became a standard reagent in alchemical and medical texts.

The principles of controlled heating and distillation used by Jābir ibn Hayyān and his successors influenced the development of laboratory techniques that are still in use. Sulfuric acid eventually became a cornerstone of the chemical industry during the Industrial Revolution, used in the production of fertilizers, dyes, and other essential chemicals.

The early production of sulfuric acid by Iranian alchemists marked a critical advancement in the history of chemistry. Through careful experimentation and dedication to understanding the natural world, scholars like Jābir ibn Hayyān paved the way for modern scientific practices and industrial applications of chemistry. Their work highlights how ancient Persia played a foundational role in the global history of science, demonstrating that the roots of modern chemistry run deep into the past. The legacy of their discovery lives on today, as sulfuric acid remains one of the most widely used chemicals in industry and research.

The story of the invention of the Qanat system

The qanat system is one of ancient Iran's most remarkable engineering achievements, designed to provide sustainable and reliable water supplies in arid regions. Developed over 3,000 years ago, this underground water management system enabled communities to access fresh water from deep aquifers, turning dry and desert landscapes into fertile land suitable for agriculture and settlement. The qanat system not only transformed life in ancient Iran but also spread across the Middle East, North Africa, and parts of Europe, leaving a lasting legacy in water engineering.

Origins of the Qanat System

The exact origins of the qanat system are hard to pinpoint, but most historians believe it was invented in ancient Persia (modern-day Iran) around the 1st millennium BCE, possibly during the reign of the Achaemenid Empire (550–330 BCE). The system may have originated in the regions of Zagros Mountains or the Khorasan desert, where water resources were scarce, and people were forced to innovate to ensure a steady water supply.

Due to the challenges posed by the Iranian plateau's dry climate, where surface water quickly evaporates, the ancient Persians devised a system that would tap into underground water sources while minimizing evaporation. They created the qanat system, a series of gently sloping, underground tunnels that transported groundwater from the foothills of mountains to lower-lying, more arid areas.

How the Qanat System Works

The qanat system relies on gravity to transport water from high-altitude aquifers to the surface without the need for pumping. It consists of three main components:

1. **Mother Well**: The qanat starts with a deep vertical shaft known as the mother well, which is dug down to the water table, typically located in a mountainous region. This well taps into an underground aquifer, providing a reliable source of water.

2. **Underground Tunnel (Qanat)**: From the mother well, a gently sloping underground tunnel is dug, often extending several kilometers, leading to the area where the water is needed. The slope is engineered carefully to allow water to flow by gravity alone. These tunnels were usually dug by skilled workers called "muqannis," who would work in extreme conditions to complete the system.

3. **Vertical Shaft (Air and Access Shafts)**: Along the length of the qanat tunnel, vertical shafts are dug at intervals to allow for ventilation and access. These shafts also serve to remove excavated soil, and they help stabilize the tunnel structure. In case of repairs, the shafts provide access points for workers.

At the end of the tunnel, the water emerges at the surface in a distribution basin, where it can then be used for irrigation, drinking, and other purposes.

The Engineering Genius Behind Qanats

Building a qanat required remarkable engineering skills, knowledge of hydrology, and an understanding of geology. The design took advantage of natural gradients and required precise calculations to ensure a steady flow without any flooding. The gradual slope and underground channels minimized water loss from evaporation, which is essential in Iran's hot and arid regions. Furthermore, the durability of qanats is impressive—some qanat systems have been continuously in use for over a thousand years.

The muqannis who dug qanats were highly skilled and respected artisans. They passed down their knowledge through generations and were able to determine the depth and quality of underground water sources based on the local geography and signs like the type of vegetation and soil. The success of a qanat system depended heavily on the skill and experience of these workers, making them indispensable members of their communities.

***

The success of the qanat system in Persia inspired other civilizations to adopt and adapt it to their own landscapes. The technology spread to neighboring regions, including Mesopotamia, Egypt, Arabia, and North Africa, and eventually reached Spain during the Islamic period. From Spain, it spread to other parts of Europe and influenced water management practices around the world.

In many places, qanat systems remain in use today, especially in Iran, where these ancient aqueducts continue to supply water to farms and communities. In regions like North Africa and the Middle East, modern engineering has sometimes replaced traditional qanats, but the principles of the qanat system still inform sustainable water management.

The story of the innovation of Algebra

The invention and development of algebra by Persian scholars, particularly during the Islamic Golden Age, represents a pivotal moment in the history of mathematics. Algebra is a branch of mathematics that deals with equations, variables, and their relationships. This field owes much of its origins and early development to the work of Muhammad ibn Musa Khwarizmi, a Persian mathematician whose groundbreaking contributions laid the foundation for modern algebra. His work was not only essential to mathematical progress in the Islamic world but also profoundly influenced European mathematics and education centuries later.

Background: Mathematics in the Ancient World

Before Khwarizmi, the use of mathematics and arithmetic was widespread in ancient civilizations like those of Babylon, Egypt, and Greece. However, these early forms of mathematics were primarily concerned with geometry, practical calculations, and specific problems rather than general principles and theories. The Babylonians and Egyptians used arithmetic for practical purposes, like measuring land and constructing buildings, but they lacked a formalized, symbolic way of solving equations.

The Greek mathematicians, especially Euclid and Diophantus, advanced mathematics significantly, but their work primarily focused on geometry and the theory of numbers. Diophantus's "Arithmetica" was one of the few works that dealt with solutions to equations, though in a much less systematic form. These early mathematical ideas laid the groundwork, but it was Khwarizmi who brought a new approach to the field, creating an organized method for solving equations that evolved into the discipline we now call algebra.

Muhammad ibn Musa Khwarizmi: The Father of Algebra

Khwarizmi's most influential work, "Kitab Mukhtasar fi Hisab al-Jabr wal-Muqabala" (The Compendious Book on Calculation by Completion and Balancing), written around 820 CE, is considered the first systematic text on algebra. The word "algebra" itself is derived from "al-Jabr," one of the operations Khwarizmi used to solve equations, which roughly translates to "restoration" or "completion." In this work, Khwarizmi presented methods for solving linear and quadratic equations systematically, using both words and arithmetic methods.

Key Concepts in Khwarizmi's Algebra

Khwarizmi's "al-Jabr" introduced a new way of thinking about mathematics. Unlike the Greek tradition, which primarily relied on geometry, Khwarizmi's approach focused on symbols and operations, making it possible to work with unknown quantities and generalize solutions to equations. Some of his key contributions include:

1. **Methods for Solving Equations**: Khwarizmi explained methods for solving both linear and quadratic equations. He did so without symbolic notation, using words and operations to describe the steps. For example, he provided solutions for equations like $ax^2 + bx = c$ by completing the square, a technique still taught in algebra today.

2. **Operations of Al-Jabr and Al-Muqabala**: Al-Jabr (completion) involved moving terms from one side of an equation to another to simplify it, while al-Muqabala (balancing) involved combining like terms. These principles laid the groundwork for operations on equations that are essential in algebra.

3. **Practical Applications**: Unlike purely theoretical mathematics, Khwarizmi's book included practical problems relating to inheritance law, commerce, and surveying, showing how algebra could be applied to real-life situations. This practical emphasis made his work accessible and relevant to many people beyond the academic world.

4. **A Systematic Approach**: Khwarizmi's work represents the first known systematic study of algebraic operations, treating it as an independent discipline rather than a set of unrelated rules for solving problems.

The Spread and Influence of Khwarizmi's Algebra

Khwarizmi's works were translated into Latin during the 12th century, becoming foundational texts in European education. His book on algebra was translated as "Algoritmi de Numero Indorum" in Europe, leading to the adoption of the term "algorithm" from his name, which today denotes any step-by-step problem-solving method. His Latin translations, especially of "al-Jabr," made him widely known among medieval scholars, bringing the principles of algebra to Europe and shaping the future of mathematics.

The significance of Khwarizmi's work on algebra lies in how it unified and generalized mathematical knowledge. His methods enabled mathematicians to solve equations without relying on geometrical constructions. The abstract principles he developed laid the foundation for future generations to advance algebra, eventually leading to the symbolic notation and advanced equations used today.

Further Development by Persian Scholars

Following Khwarizmi, other Persian scholars continued to refine and expand algebra. Omar Khayyam, a Persian polymath, made significant contributions in the 11th century by solving cubic equations using geometric methods and attempting to classify equations systematically. Sharafaddin Tusi, another Persian mathematician, explored polynomial equations in the 12th century and made advances that anticipated some concepts in modern algebra.

The work of these scholars reflected a thriving culture of learning and intellectual curiosity within the Islamic Golden Age. They helped preserve and expand on Greek mathematical knowledge, integrating it with innovations from India and Persia. As algebra grew as a discipline, its abstract principles became a crucial foundation for other scientific and mathematical developments.

Importance of Algebra and Its Legacy

Algebra's impact on mathematics and science has been profound. The discipline provides a foundation for a vast array of fields, including engineering, physics, computer science, and economics. It enables abstract thinking, modeling complex systems, and solving problems in fields from finance to physics. Key areas influenced by algebra include:

1. **Modern Mathematics and Calculus**: The methods Khwarizmi and his successors introduced are essential in solving equations and handling variables. Algebra directly influenced the development of calculus and other advanced mathematical disciplines.

2. **Science and Engineering**: Algebra allows for the formulation of mathematical models that describe physical, chemical, and biological processes. This has been vital in fields like engineering, where algebraic methods help analyze and design systems.

3. **Computer Science and Cryptography**: Algebraic principles underpin algorithms, coding theory, and data encryption, which are foundational to computer science and data security.

4. **Education**: Algebra is now a cornerstone of education worldwide. Its focus on logical reasoning and problem-solving skills has made it an essential subject, shaping the way students approach complex ideas.

***

The invention and development of algebra by Persian scholars, particularly by Muhammad ibn Musa Khwarizmi, marks a transformative period in the history of mathematics. Khwarizmi's pioneering work "al-Jabr" laid the foundations for a universal language of equations and variables that has empowered countless discoveries and innovations. By introducing a structured approach to solving equations and making mathematics accessible and applicable to real-world problems, Khwarizmi's contributions ensured that algebra would become a permanent fixture in human knowledge. His legacy, along with that of other Persian mathematicians, lives on in classrooms, research labs, and industries around the world, a testament to the lasting impact of his revolutionary work in mathematics.

The story of the invention of Alcohol extraction

The invention of alcohol extraction is a significant achievement attributed to early Persian scientists and alchemists. This breakthrough, which involved the distillation of liquids to isolate ethanol, or "pure alcohol," had profound implications for chemistry, medicine, and numerous fields beyond. The process of distilling and refining alcoholic substances was developed and refined in Persia during the Islamic Golden Age, especially through the work of pioneering scholars like Razi (Rhazes), who advanced the method and laid the foundations for distillation techniques still in use today.

Background: Early Alchemy and Distillation in Persia

One key area of study was the distillation of liquids, initially to create perfumes, essential oils, and medicines. Distillation is the process of heating a liquid to create vapor and then condensing it back into a liquid. By separating liquids with different boiling points, early chemists realized they could isolate specific components. Distillation methods had been practiced in Egypt and Mesopotamia, but it was the Persian scholars who refined the process to isolate "pure" alcohol.

Razi (Rhazes) and the Refinement of Distillation

One of the most influential figures in the development of alcohol extraction was Muhammad ibn Zakariya Razi (865–925 CE), commonly known in the West as Rhazes. A Persian polymath, physician, and alchemist, Razi was dedicated to advancing medicine and chemistry. He was among the first to refine the distillation process to produce a concentrated form of alcohol, which he referred to as **al-kohl** (from which the word "alcohol" is derived).

Razi wrote extensively on his methods in texts such as **"Kitab al-Asrar"** (The Book of Secrets) and **"Kitab al-Mansuri"** (The Book for al-Mansur). In these works, he detailed the processes for distilling various substances, including wine, to create purified alcohol, which he described as having both medicinal and chemical uses.

The Distillation Process: How Alcohol Was Extracted

The method of alcohol extraction employed by Razi and other Persian alchemists involved a multi-step distillation process:

1. **Heating the Liquid**: Wine or fermented fruit juice would be heated in a closed distillation vessel. As the temperature increased, the alcohol vaporized before the water due to its lower boiling point.
2. **Collecting the Vapor**: The alcohol vapor rose into a cooling tube or an alembic. The tube was designed to separate alcohol vapors from other components.
3. **Condensation and Collection**: As the vapor cooled, it condensed back into liquid form and was collected in a separate vessel, now containing a higher concentration of alcohol. Repeated distillations could increase the purity further, creating a "pure" form of ethanol.

The distillation method required precision, as it was necessary to control temperature to separate alcohol effectively from water and other components. Razi's distillation apparatus, similar to a modern alembic, was sophisticated for its time, reflecting the ingenuity of Persian alchemists in creating complex chemical processes with limited technology.

Applications and Importance of Alcohol Extraction

The invention of alcohol extraction had several significant applications and benefits:

1. **Medicinal Use**: Concentrated alcohol became an essential component in medicine. Persian physicians used it as a solvent for medicinal herbs, creating tinctures and disinfecting wounds. Alcohol's antiseptic properties were especially valued in wound treatment and surgical procedures.

2. **Alchemy and Chemistry**: Alcohol extraction allowed alchemists to experiment with pure substances, advancing their understanding of chemical reactions. Isolating alcohol as a unique substance helped differentiate between compounds and inspired further research into other elements and compounds.

3. **Perfume and Essence Production**: Distillation was also used to produce perfumes, an important industry in the Islamic world. Alcohol extraction allowed alchemists to isolate and concentrate fragrances from flowers and herbs, producing perfumes that were prized across empires.

4. **Alcohol as a Catalyst**: Alchemists found that alcohol could be used to dissolve and purify other substances, which was invaluable in various alchemical experiments. This finding laid the groundwork for understanding solvents, leading to advancements in chemistry.

5. **Scientific and Cultural Exchange**: Persian scholars documented their findings, which were later translated into Latin and shared with European scholars during the Middle Ages. The knowledge of distillation and alcohol extraction became part of European chemistry and medical science, influencing scholars and scientists like Paracelsus.

Influence on Global Chemistry and Medicine

The process of alcohol extraction pioneered by Persian alchemists had a profound impact beyond the Islamic world. By the 12th century, Latin translations of works by Razi and other Persian alchemists spread to Europe, where distillation and alchemical knowledge merged with local practices. This exchange influenced the development of early European pharmacology and chemistry, as distillation became a common method for isolating medicinal compounds and other substances.

As Renaissance scholars like Paracelsus explored alchemy, they refined distillation further, using it to produce stronger medicinal tinctures and other compounds. The development of chemistry as a scientific discipline owes much to these early methods, which began with Persian alchemists' experiments in isolation and purification.

The invention of alcohol extraction by Persian scholars, especially through the work of Razi, represents a landmark in the history of chemistry and medicine. By refining the distillation process to isolate ethanol, Persian alchemists laid the foundations for centuries of scientific progress. The methods they developed for purifying and isolating substances allowed for the emergence of pharmacology and more precise chemical analysis, forming the basis of laboratory techniques that are still essential today.

Persian scholars' legacy in alcohol extraction and distillation not only advanced science in the Islamic world but also facilitated the global exchange of knowledge, influencing science and medicine in Europe and beyond. Their achievements stand as a testament to the enduring impact of early Islamic science on the modern world.

The story of the Gondishapur University

The University of Gondishapur (also spelled Gundeshapur), founded in the 3rd century CE, is one of the world's oldest known universities and research institutions, situated in the ancient Persian city of Gondishapur (near present-day Ahvaz, Iran). Established during the Sassanian Empire by Shah Shapur I and reaching its peak under Khosrow I (Anushirvan), this institution served as a leading center for higher education and scientific research. The academy was notable for integrating knowledge from diverse cultural traditions and setting a model for future universities, influencing both Islamic and Western education for centuries.

Founding and Early History

The story of Gondishapur begins with Shapur I (241–272 CE), a Sassanian emperor who founded the city as a stronghold in southwest Persia. Initially, Gondishapur was primarily a place for political governance and military planning. However, its location in the fertile region of Khuzestan made it a strategic and cultural crossroads, where Persian, Greek, Indian, and later Roman and Christian influences converged.

Over time, Gondishapur evolved from a fortified city into a hub for scholarly pursuits. It was during the reign of Khosrow I (531–579 CE), known for his passion for learning and cultural exchange, that Gondishapur flourished as an educational center. Khosrow I's support for scholars of diverse backgrounds established Gondishapur as a leading academy for learning, scientific investigation, and medical training, setting a foundation for the concept of a university.

The Academy's Structure and Multicultural Legacy

The Academy of Gondishapur was a truly multicultural institution, with scholars from various backgrounds, including Greek, Indian, Persian, and Roman, who contributed to fields such as medicine, astronomy, philosophy, and mathematics. The academy attracted many scholars fleeing persecution or seeking patronage, including Nestorian Christians from the Byzantine Empire who brought with them knowledge of Greek philosophy and science.

The academy's structure was unique for its time. It functioned as:

1. **A Medical and Teaching Hospital**: Gondishapur's hospital was among the earliest examples of a teaching hospital. Students trained under the guidance of experienced physicians, observing and learning through practical medical experience. This model of combining education with clinical practice became standard in future medical schools worldwide.

2. **A Research Institution**: The scholars at Gondishapur conducted research in various fields. It was a place where scholars could freely engage in intellectual pursuits, experiment, and record their findings, forming the basis of academic research practices seen in later institutions.

3. **A Translation Center**: Gondishapur was also an intellectual bridge, translating and preserving texts from Greek, Sanskrit, and Syriac into Pahlavi (Middle Persian). This exchange enabled the diffusion of Greek philosophy, Indian medicine, and Persian scientific

knowledge, enriching the curriculum and providing a foundation for future Islamic scholarship.

Contributions to Medicine, Mathematics, and Astronomy

Gondishapur was particularly renowned for its contributions to medicine. Physicians at the academy studied and advanced practices in surgery, pharmacology, and holistic health. They compiled medical texts and contributed to the medical canon by integrating Greek medicine, such as the works of Hippocrates and Galen, with Persian and Indian medical traditions, creating a more comprehensive understanding of health and disease.

Some notable fields and achievements include:

- **Medicine and Surgery**: The academy trained physicians who excelled in diagnosing illnesses and performing surgeries. Gondishapur's influence on medicine extended throughout the Islamic Golden Age, inspiring future Persian practitioners like Razi and Ibn Sina (Avicenna).

- **Mathematics**: Scholars in Gondishapur contributed to developments in mathematics, including the use of the decimal system and advances in algebra and geometry. These foundations were crucial for the later achievements of Islamic mathematicians.

- **Astronomy and Philosophy**: The academy promoted studies in astronomy and natural philosophy, fields that were enriched by Indian and Greek knowledge. Scholars at Gondishapur created astronomical tables and observed celestial phenomena, fostering an early scientific understanding of the cosmos.

***

The Academy of Gondishapur stands as one of the earliest examples of a university, combining research, medical training, cultural exchange, and the translation of scientific knowledge. It was a place where Persian emperors, valuing wisdom and knowledge, nurtured intellectual growth and set the foundation for future centers of learning. Gondishapur's contributions to medicine, philosophy, and scientific inquiry enriched the Islamic world and, through it, medieval Europe, ultimately leaving a legacy that shaped the course of human knowledge.

Day 9: Turkey

Turkey has made significant contributions across various fields, from science and technology to the arts, architecture, and culinary traditions. Here are fifteen of the most impactful innovations, explorations, and inventions introduced by Turks and the broader Ottoman Empire:

1. **Surgical Instruments by Al-Zahrawi (Albucasis)**: Although born in Andalusia, Al-Zahrawi, a prominent figure in Islamic medicine, influenced Ottoman surgical practices. Turkish doctors used and expanded upon his advanced surgical tools and methods, some of which are still used in modified forms today.

2. **Coffee Culture**: Coffee was first introduced to the Ottoman Empire from Yemen in the 16th century and quickly became central to Turkish social life. Coffeehouses became widespread in Istanbul and later spread to Europe, introducing the world to coffee culture as we know it today.

3. **Tulip Cultivation and the "Tulip Era"**: The 18th-century Ottoman "Tulip Era" saw an explosion of interest in tulip cultivation, influencing art, fashion, and architecture. Tulips became synonymous with elegance in European gardens and heavily influenced Dutch horticulture.

4. **The Use of Gunpowder in Cannons**: The Ottomans were among the first to use gunpowder extensively in warfare, particularly with large cannons, like the ones used in the conquest of Constantinople in 1453. This influenced military tactics and cannon technology in Europe.

5. **Vaccination Practices**: Ottoman practices in inoculation for smallpox influenced the development of vaccinations. Lady Mary Wortley Montagu, the wife of the British ambassador to the Ottoman Empire, brought back the practice to England, helping inspire the modern vaccine movement.

6. **Astrolabe Advancements**: Ottoman astronomers and navigators made refinements to the astrolabe, an ancient instrument for determining latitude. These improvements supported Islamic and European navigation and astronomy.

7. **Architectural Domes and the Influence of Mimar Sinan**: Architect Mimar Sinan developed advanced techniques for building large domes, influencing architecture worldwide. His designs for mosques, such as the Suleymaniye Mosque, inspired structures including the Taj Mahal and the Blue Mosque.

8. **The Istanbul Observatory**: In the 16th century, Taqi al-Din, an Ottoman astronomer, established an advanced observatory in Istanbul. His contributions to astronomy, including observations and calculations, influenced both Islamic and European astronomy.

9. **Tulumba Pump (Water Pump)**: In the Ottoman Empire, a tulumba (a type of reciprocating water pump) was widely used for irrigation, firefighting, and supplying water in urban centers, showcasing early hydraulic engineering techniques.

10. **Turkish Baths (Hammams)**: Although inspired by Roman baths, Turkish hammams became unique social and architectural spaces central to Turkish culture. The hammam spread across the Ottoman Empire and influenced bathhouse designs in many parts of Europe and the Middle East.

11. **Catapults and Siege Techniques**: Ottoman engineers developed advanced siege weapons and techniques that were crucial in their expansion. They refined the use of catapults, battering rams, and other technologies to lay successful sieges in Europe, Asia, and North Africa.

12. **Lokum (Turkish Delight)**: Lokum, commonly known as Turkish Delight, was invented in the Ottoman Empire and became a popular sweet across Europe. Its unique taste and texture have inspired variations and other confections globally.

13. **Military March Music (Mehter)**: The Mehter band, considered one of the first military bands, was established in the Ottoman Empire to boost troop morale. This tradition influenced the formation of military bands in Europe, impacting classical music and military traditions.

14. **Culinary Innovations (Kebabs and Baklava)**: Turkish culinary contributions include popular dishes like kebabs and baklava, which spread widely through the Ottoman Empire and became

iconic foods worldwide. These foods remain central to many Middle Eastern and Mediterranean cuisines today.

These contributions from Turkey and the Ottoman Empire reflect a legacy of cultural, scientific, and artistic advancements that continue to influence modern life globally. From coffee culture and culinary delights to military innovations and architectural masterpieces, these contributions showcase the significant impact of Turkish civilization.

The story of the innovation of the Vaccination

The innovation of vaccination practices in Turkey has its roots in the early adoption and advancement of immunization techniques, particularly during the time of the Ottoman Empire. The development of vaccination practices in the region can be traced back to variolation—an ancient method of inoculating individuals with small amounts of smallpox virus to induce immunity, long before the discovery of modern vaccines. This practice was crucial in the eventual development of vaccination as we know it today, and Turkish doctors played a key role in advancing this practice.

Early History: The Practice of Variolation

The concept of immunization through variolation (also called inoculation) dates back to ancient times in various cultures, including China and India. Variolation involved deliberately infecting an individual with a mild form of smallpox (often through scratching the skin) in hopes that the person would develop immunity to the disease. This practice was not completely safe, but it was a precursor to modern vaccination techniques.

In the Ottoman Empire, variolation began to be practiced as early as the 17th century. Turkish doctors, particularly in Istanbul, played a significant role in refining and popularizing this technique. Historical accounts suggest that the Ottoman sultans and scholars were well aware of the practice's potential to protect people from smallpox.

The Ottoman Empire was a key bridge in the spread of medical knowledge, and it facilitated the exchange of knowledge between East and West. While the practice of variolation was common in the Ottoman

Empire, it spread to Europe in the early 18th century through the efforts of Lady Mary Wortley Montagu, the wife of the British ambassador to the Ottoman Empire.

In 1717, Lady Montagu visited Istanbul and witnessed variolation firsthand. She was impressed by its effectiveness and even had her own children inoculated. She introduced the concept to England upon her return, and variolation quickly spread throughout Europe. The practice gained popularity in England, France, and other parts of Europe, laying the groundwork for the eventual development of the smallpox vaccine.

The Introduction of Vaccination: Edward Jenner and the Ottoman Connection

Although Edward Jenner, an English physician, is credited with the discovery of the modern smallpox vaccine in 1796, his work was indirectly influenced by practices in Turkey. Jenner's observation that cowpox (a disease affecting cows) could confer immunity to smallpox was groundbreaking. However, it is believed that Jenner's concept of vaccination was inspired by the earlier practice of variolation, which had already been practiced in the Ottoman Empire for centuries.

The Ottoman connection to Jenner's work became clearer when it was discovered that Turkish physicians had been practicing smallpox inoculation long before Jenner. Jenner himself acknowledged that his ideas were influenced by the global knowledge of variolation, including practices that had originated in Turkey. It's believed that Turkish doctors had been using cowpox as an alternative method of inoculation in some regions, possibly even before Jenner's discovery, although there is little documented evidence of this.

Day 10: Greece

Ancient Greece has been instrumental in shaping many aspects of Western civilization, with contributions in philosophy, science, mathematics, government, and the arts. Here are fifteen of the most important innovations, explorations, and inventions introduced by the Greeks:

1. **Democracy**: Ancient Athens developed one of the first known forms of democracy around the 5th century BCE. This Athenian system allowed citizens to participate directly in decision-making, inspiring modern democratic governments worldwide.

2. **Philosophy**: Greek philosophers such as Socrates, Plato, and Aristotle laid the foundations of Western philosophy. Their work influenced ethics, metaphysics, logic, and political theory, and remains central to philosophy curricula today.

3. **Geometry and Mathematics**: Greek mathematicians like Euclid, Pythagoras, and Archimedes made groundbreaking contributions to geometry, algebra, and calculus. Euclid's *Elements* became one of the most influential works on mathematics.

4. **Medicine**: Hippocrates, known as the "Father of Medicine," established medicine as a distinct field based on observation and rationality. The Hippocratic Oath, originating from his school, continues to be a foundational ethical standard in medicine.

5. **Theater and Drama**: Ancient Greeks invented dramatic genres like tragedy and comedy. Playwrights like Aeschylus, Sophocles, and Aristophanes laid the groundwork for Western theater, influencing storytelling, stagecraft, and performance art.

6. **The Olympics**: The ancient Olympic Games began in Olympia in 776 BCE as a religious festival honoring Zeus. The concept of competitive athletic games in the spirit of peace and honor was revived in the 19th century, inspiring the modern Olympic Games.

7. **Astronomy**: Greek astronomers like Hipparchus and Ptolemy made significant contributions to early astronomy. Hipparchus developed one of the first star catalogs and calculated the Earth's

precession, while Ptolemy's *Almagest* was an authoritative text in astronomy for centuries.

8. **Mythology and Epic Poetry**: Greek mythology and epic poetry, especially through works like Homer's *Iliad* and *Odyssey*, have deeply influenced Western literature, art, and culture. These stories provide insights into ancient Greek values and continue to inspire modern storytelling.

9. **Historiography**: Greeks like Herodotus and Thucydides pioneered the recording and study of history. Their works emphasized evidence-based analysis and narrative structure, influencing how historical records are maintained and interpreted today.

10. **Democratic Government Architecture (Agora)**: The Greek concept of the *agora*, or public space, served as a marketplace and a forum for political discussions, forming the foundation of public engagement in civic life, a concept seen in many democratic governments.

11. **The Water Mill**: Greeks are credited with inventing the first water mills around the 3rd century BCE. These mills used waterpower to grind grain and perform other tasks, a major advancement in energy technology that spread through the Roman Empire and beyond.

12. **Lighthouses (The Lighthouse of Alexandria)**: The Lighthouse of Alexandria, also known as the Pharos of Alexandria, was one of the Seven Wonders of the Ancient World and one of the first large-scale lighthouses. It set a precedent for coastal navigation that is essential to modern maritime trade.

13. **The Archimedes Screw**: Invented by Archimedes, this screw-like device was used to lift water for irrigation and other purposes. It remains one of the fundamental principles in fluid mechanics and is still used in various applications today.

14. **The Development of Logic and Rhetoric**: Aristotle's work in logic and rhetoric laid the foundations for Western intellectual thought. His studies in syllogism and persuasive speaking are still integral to fields like law, politics, and philosophy.

15. **Engineering Marvels (Antikythera Mechanism)**: Discovered off the coast of Greece, the Antikythera mechanism is an ancient

analog computer dating back to 150–100 BCE. It was used to predict astronomical positions and eclipses, showcasing the Greeks' advanced understanding of engineering and astronomy.

These contributions from the Greeks have had a lasting impact on diverse fields and continue to be relevant in modern society, providing a basis for Western philosophy, science, politics, and culture.

The story of the invention of Democracy

The invention of democracy is widely credited to ancient Greece, particularly the city-state of Athens during the 5th century BCE. This development marked a significant shift in the governance of societies, where power was vested in the people rather than in a monarch, oligarchs, or an aristocracy. The origins of Greek democracy are complex, and its evolution reflects both revolutionary ideas and practical experimentation in how to govern large, diverse populations. Athens is traditionally seen as the birthplace of democracy, and its political system laid the groundwork for democratic ideas that would later influence the formation of modern democratic governments.

Background: Political Structure Before Democracy

Before the invention of democracy, Greek city-states, or polis, were governed by kings, aristocrats, or tyrants. These rulers held power without the consent of the people, and the general populace had limited political rights. In Athens, for example, during the Archaic period (circa 800–500 BCE), the city-state was ruled by an oligarchy—a small group of wealthy and powerful citizens who made decisions for the state without consulting the majority.

Athens, like many Greek city-states, was initially ruled by tyrants in the 7th and 6th centuries BCE. Tyrants were not necessarily cruel or despotic rulers; they were simply individuals who seized power, often with the support of the common people, by overthrowing the existing aristocratic elites. However, these tyrants still maintained autocratic control over the state, which led to growing dissatisfaction among the citizens.

Early Steps Toward Democracy: The Reforms of Solon

The development of democracy in Athens was not a sudden event but rather the result of a series of reforms and political changes that gradually expanded the participation of citizens in government.

The first major step toward democratization came with the reforms of Solon in the early 6th century BCE. Solon was a statesman, poet, and lawgiver who was appointed to mediate between the aristocracy and the common people. He introduced a series of legal reforms designed to alleviate the social and economic pressures on the lower classes and curb the power of the aristocracy.

- **Abolition of Debt Slavery**: Solon famously abolished debt slavery, which had plagued the poorer citizens of Athens. Many Athenians had been forced into slavery because of unpaid debts to aristocratic landowners. Solon's reforms allowed for the redistribution of land and wealth, which helped to ease tensions between the classes.

- **Creation of New Laws**: Solon also created a new body of laws, which were written down and made publicly available, ensuring that no one could rule arbitrarily or change laws at will. He introduced a system of courts where citizens could appeal to a jury of their peers, making justice more accessible.

- **Expansion of Citizenship**: Solon's reforms laid the groundwork for greater political participation, though his reforms did not create a fully democratic system. His policies still limited political participation to a select group of citizens based on wealth and status.

Despite Solon's reforms, Athens still remained an oligarchy. The rich and powerful families continued to wield significant influence over the political process. However, his reforms set the stage for the next major step in the development of democracy.

The Rise of Cleisthenes and the Foundation of Democracy

The true birth of Athenian democracy is often attributed to the statesman Cleisthenes, who enacted a series of radical reforms in 508 BCE that fundamentally reshaped Athens' political system. Cleisthenes, known as the "Father of Athenian Democracy," introduced changes that empowered ordinary citizens and created a more inclusive government.

- **Reorganization of the Athenian Tribes**: Cleisthenes reorganized the Athenian population into ten tribes, based not on family heritage but on geographic location. This restructuring diluted the power of the aristocratic families and allowed more citizens to participate in the political process.

- **Creation of the Council of 500**: Cleisthenes also established the Council of 500, which replaced the previous aristocratic council. The Council was responsible for preparing the agenda for the Assembly (the main governing body of Athens). Unlike earlier councils, membership in the Council was determined by a lottery, ensuring a more democratic and equal representation of the citizenry.

- **Direct Participation in Government**: Cleisthenes' reforms gave ordinary male citizens the right to participate directly in political decisions. The central governing body of Athens was the Assembly (Ekklesia), which was open to all male citizens of Athens over the age of 18. The Assembly had the power to make decisions on important matters such as war, foreign policy, and legislation.

- **Use of Ostracism**: To prevent any individual from gaining too much power, Cleisthenes introduced the practice of ostracism. Every year, the citizens of Athens could vote to exile a person for ten years if they were perceived as a threat to the democratic system. Although ostracism was rarely used, it served as a safeguard against tyranny.

These reforms significantly increased the political power of the average citizen, and they were instrumental in creating the democratic system that became a model for future governments.

The Democratic System in Action

With the reforms of Cleisthenes, Athens became a direct democracy, where eligible male citizens (about 30–40% of the population) could participate in political decision-making. This system was radically different from the oligarchies and monarchies that existed in other Greek city-states.

The democratic process in Athens worked through the following mechanisms:

1. **The Assembly (Ekklesia)**: The Assembly was the central body of Athenian democracy, where all citizens could vote on important issues. Meetings were held regularly, and any citizen could speak and propose laws or decisions. The Assembly's power extended to making decisions on matters of war, foreign policy, the economy, and public projects.

2. **The Council of 500**: The Council of 500 prepared the agenda for the Assembly and ensured that the democratic process ran smoothly. Members of the Council were selected by a lottery and served for one year.

3. **The Courts**: Athens had a system of popular courts in which large juries of citizens (often hundreds) would decide legal cases. The courts were an essential part of Athenian democracy, as they allowed citizens to participate in legal matters and ensured that justice was administered fairly.

4. **Political Equality**: Athenian democracy emphasized political equality (isonomia), where every eligible citizen had the right to participate in the political process. This was revolutionary in an era where political power was typically reserved for elites.

The Limitations of Athenian Democracy

While Athens is often celebrated for its pioneering form of democracy, it was far from perfect. Key limitations of the system included:

- **Exclusion of Women and Slaves**: Athenian democracy was limited to free male citizens, meaning that women, slaves, and metics (foreigners living in Athens) were excluded from political participation. This meant that only a small portion of the population had the right to participate in democratic processes.

- **Direct Democracy**: While direct democracy allowed for widespread participation, it also meant that important decisions could be made impulsively by the citizens, sometimes leading to hasty or ill-considered policies.

The Influence of Athenian Democracy on Modern Democracy

Despite its limitations, Athenian democracy served as a model for democratic ideas in later civilizations. The principles of participatory government, citizen involvement, and accountability laid the foundation for modern democratic systems. The Roman Republic, the Enlightenment philosophers such as John Locke and Jean-Jacques Rousseau, and the American and French Revolutions all drew inspiration from the concepts of democracy first practiced in Athens.

***

The invention of democracy in ancient Greece, particularly in Athens, represents a monumental shift in how societies could be governed. Through the reforms of Solon, Cleisthenes, and others, Athens created a system where the power of government was vested in its citizens, allowing for a more inclusive and participatory form of rule than had existed before. While Athenian democracy was far from perfect, its legacy lives on today in democratic governments around the world.

The story of the invention of the Olympic Games

The story of the Olympic Games traces its origins to ancient Greece, where the Games were created as a religious festival honoring Zeus, the king of the Greek gods. The first recorded Olympic Games took place in 776 BCE, and they became one of the most significant events in ancient Greek culture, celebrated every four years in the city of Olympia in the western part of the Peloponnese Peninsula.

The Origins of the Olympic Games

The exact origins of the Olympic Games are steeped in myth and legend. According to one story, the Games were founded by Heracles (Hercules), the hero of Greek mythology, who established the competition as a way to celebrate his Twelve Labors and to honor Zeus. Another myth attributes the foundation of the Games to King Oenomaus of Pisa, who organized a race between his daughter and her suitors.

The Games were initially a religious festival held in honor of Zeus at his sanctuary in Olympia, a site that was considered sacred. The ancient Greeks believed that the Games not only honored Zeus but also represented a way to bring peace and unity among the often-warring Greek city-states.

The First Olympic Games (776 BCE)

The first Olympic Games, held in 776 BCE, consisted of just a single event: a foot race known as the stade race, which was approximately 192 meters long. This race was the central event of the Games and the only event for the first few centuries. Over time, more events were added, including the pentathlon, which featured five events: running, jumping, discus throw, javelin throw, and wrestling.

Expansion of Events and Inclusion of More Sports

As the Olympic Games gained popularity, more sports were introduced. By the 5th century BCE, events like boxing, pankration (a no-holds-barred form of mixed martial arts), and chariot racing were added to the Games. Competitions were held in both individual and team events, and the Games expanded to include a variety of athletic challenges.

The Role of Athletes and the Olympic Flame

In ancient Greece, athletes were highly respected, and winning an Olympic event was considered one of the greatest honors a Greek citizen could achieve. Victors were celebrated with parades, poems, and statues. In some cases, victors were even given lifelong pensions and other privileges. To compete in the Games, athletes had to undergo rigorous training and were required to swear an oath of fairness.

A key aspect of the ancient Olympic Games was the sacred flame, which was kept burning throughout the event. The torch was lit by the rays of the sun and symbolized the presence of the gods. The Olympic flame and torch relay, as we know it today, were revived during the modern Olympic Games in the 20th century.

The Abolition of the Ancient Olympic Games

The ancient Olympic Games continued to be held every four years for nearly 12 centuries. However, the Games were eventually abolished in 393 CE by the Roman Emperor Theodosius I, who sought to suppress pagan practices and promote Christianity throughout the Roman Empire. This marked the end of the ancient Olympic tradition, and the Games lay dormant for over 1,500 years.

The Revival of the Olympic Games (Modern Era)

In the late 19th century, there was growing interest in reviving the Olympic Games, especially with the rise of the modern spirit of nationalism and international competition. In 1896, the first modern Olympic Games were held in Athens, Greece, under the leadership of French educator Pierre de Coubertin. The Games were designed to revive the ancient ideals of athleticism, competition, and peace, and they have been held every four years since, with the exception of World Wars I and II.

The modern Olympics have expanded significantly from their ancient counterparts, now featuring a wide range of sports and athletes from around the world. The Games have evolved into one of the most prestigious and widely recognized sporting events in the world, with countries from across the globe coming together to celebrate athletic achievement and international unity.

***

The invention of the Olympic Games in ancient Greece is a story of religious devotion, athletic competition, and cultural unity. What began as a local festival in Olympia grew to become a symbol of ancient Greek culture, and through the efforts of figures like Pierre de Coubertin, the Olympic Games were revived in the modern era. Today, the Olympics continue to inspire athletes and audiences alike, drawing from the rich legacy of their ancient Greek origins.

The story of the invention of the Lighthouse

The lighthouse as we know it today has its roots in ancient Greece, with one of the most famous early examples being the Pharos of Alexandria, which became one of the Seven Wonders of the Ancient World. The invention and development of the lighthouse were closely tied to the need for guiding ships safely into harbors, particularly in the Mediterranean, where maritime trade and naval power were vital.

The Origins of the Lighthouse Concept

Lighthouses, in their earliest form, were structures built to aid in maritime navigation by making dangerous coastlines, rocks, and harbors visible at night or in bad weather. Ancient cultures like the Egyptians, Phoenicians, and Greeks used various types of beacon fires to guide ships, but it was the Greeks who are often credited with creating the first truly monumental lighthouse.

In ancient Greece, sailors would light large signal fires on hilltops or along coastlines to warn approaching ships of dangers. However, these early "lighthouses" were simple beacons that lacked the sophistication and permanence of later structures.

The Pharos of Alexandria (circa 280 BCE)

The true invention of the lighthouse is often attributed to the Greeks, specifically to the construction of the Pharos of Alexandria, one of the most impressive and influential lighthouses in history. The Pharos was built on the island of Pharos, connected by a causeway to the city of Alexandria in Egypt. It was commissioned by Ptolemy II Philadelphus, the ruler of Egypt, around 280 BCE.

The Pharos was designed by the Greek architect Sostratus of Cnidus, and it was intended to guide sailors safely into the busy harbor of Alexandria, one of the most important ports of the ancient world. The lighthouse stood at an estimated height of 100-130 meters (330–430 feet), making it one of the tallest man-made structures of the ancient world.

The Design and Function of the Pharos

The Pharos of Alexandria was an engineering marvel. It was constructed with a large tower and a flame at the top, which served as a beacon visible to ships miles away. The lighthouse was also equipped with reflective materials, such as polished bronze mirrors, which helped amplify the light, making it even more visible.

The Pharos not only served a practical purpose in guiding ships into port but also became a symbol of the strength and grandeur of the Ptolemaic Kingdom. The lighthouse was visible from great distances, making it one of the first true lighthouses in the world, as it combined both navigation and architecture in a way that had never been done before.

Legacy of the Pharos

The Pharos of Alexandria stood for over 1,500 years, surviving various earthquakes and other challenges. Eventually, it was damaged and destroyed by an earthquake in 1303 CE, but its influence lived on. The lighthouse became an iconic structure in the history of maritime navigation, and the word "pharos" itself became synonymous with the term "lighthouse."

The concept of the lighthouse spread to other parts of the ancient world, particularly in the Roman Empire, which built similar structures along important trade routes and naval hubs. The Romans continued to refine the design and function of lighthouses, using them to guide their vast fleets.

The Modern Lighthouse

The legacy of the Pharos can be seen in the development of modern lighthouses, which continue to serve as essential navigational aids for ships. The modern lighthouse has evolved from simple beacon fires to sophisticated structures with automated lights, revolving beams, and electronic signaling systems. Today, lighthouses are crucial for maritime safety and still serve as symbols of guidance and protection along coastlines.

***

The invention of the lighthouse, particularly the construction of the Pharos of Alexandria, marked a significant moment in the history of navigation and engineering. The ancient Greek development of the lighthouse concept helped shape maritime history and influenced the design of lighthouses for centuries. The Pharos, as one of the Seven Wonders of the Ancient World, remains a testament to Greek ingenuity and their contribution to modern navigation.

Day 11: Italy

Italy has a long history of groundbreaking contributions to art, science, engineering, and culture that have shaped modern civilization. Here are fifteen of the most significant innovations, explorations, and inventions introduced by Italians to the world:

1. **The Renaissance**: Italy was the birthplace of the Renaissance, a cultural movement that revolutionized art, science, literature, and philosophy. Key figures like Leonardo da Vinci, Michelangelo, and Raphael created masterpieces that changed Western art and thinking forever.

2. **Perspective in Art**: Italian artists like Filippo Brunelleschi and Leonardo da Vinci pioneered the use of linear perspective, creating a sense of depth and realism in paintings. This technique transformed Western art and is still fundamental to visual art and design.

3. **Leonardo da Vinci's Inventions**: Leonardo da Vinci designed numerous mechanical inventions, including early versions of the helicopter, parachute, and tank. His visionary engineering and sketches influenced later technological developments.

4. **Galileo's Contributions to Astronomy and Physics**: Galileo Galilei made monumental contributions to astronomy, physics, and the scientific method. His improvements to the telescope and discoveries, such as the moons of Jupiter, challenged the geocentric model and laid the groundwork for modern science.

5. **Modern Banking**: Italians, particularly the Medici family of Florence, were pioneers of modern banking during the Renaissance. They developed practices such as double-entry bookkeeping and credit, which are still used in global finance today.

6. **The Violin**: The violin, one of the most important musical instruments, was developed in 16th-century Italy by luthiers like Andrea Amati. The Cremona school of violin-making, led by Antonio Stradivari, set standards for quality that endure to this day.

7. **Opera**: Opera as an art form originated in Italy in the late 16th century. Composers like Claudio Monteverdi, Verdi, and Puccini shaped the genre, which remains a popular and influential art form worldwide.

8. **The Barometer**: Evangelista Torricelli, an Italian physicist and student of Galileo, invented the barometer in 1643. This instrument measures atmospheric pressure and was essential in developing meteorology and scientific research.

9. **The First University (University of Bologna)**: Established in 1088, the University of Bologna is considered the world's first university. It created a model for higher education that influenced universities worldwide.

10. **The Thermoscope**: Galileo Galilei invented the thermoscope, a precursor to the modern thermometer, allowing scientists to measure temperature changes. This device laid the foundation for more precise temperature measurement.

11. **Pasta and Pizza**: Italian cuisine, particularly pasta and pizza, is globally celebrated and has influenced diets around the world. Italy's innovations in culinary techniques, flavors, and ingredients continue to define international dining.

12. **Maser and Laser Technology**: Italian physicist Carlo Rubbia contributed to the development of maser technology, which later led to laser technology, widely used in medicine, telecommunications, and industry.

13. **The Electric Battery**: Alessandro Volta invented the first electric battery, the voltaic pile, in 1800. His discovery was a critical milestone in electrical science and paved the way for all modern electrical devices.

14. **Eyeglasses**: Italians are credited with inventing the first wearable eyeglasses in the 13th century. These were crucial in improving vision for millions of people and influenced optics and medical devices.

15. **Heliocentric Theory (Copernican Revolution)**: Although Polish, Nicolaus Copernicus developed his revolutionary heliocentric theory while working in Italy, challenging the Earth-centered

model and setting the stage for modern astronomy and the scientific revolution.

These Italian contributions reflect a legacy of artistic brilliance, scientific discovery, and cultural influence that continues to shape our world. From Renaissance art and opera to foundational inventions in science and engineering, Italian innovation has left an enduring global impact.

The story of the invention of Eyeglasses

The invention of eyeglasses is a fascinating story that spans several centuries, with Italy playing a pivotal role in their development. Eyeglasses, as we know them today, evolved from early attempts to improve vision, and their invention is often credited to Italian craftsmen during the late 13th century. This innovation was transformative, enabling people with vision impairments to regain the ability to read, work, and engage with the world around them. Here's the story of how eyeglasses came to be, primarily through the work of Italian inventors.

The Origins of Eyeglasses: Ancient Beginnings

Long before eyeglasses were invented, there were various tools and techniques that attempted to improve vision. The idea of magnification using lenses was known to ancient civilizations, including the Greeks and Romans. The Roman philosopher Seneca (circa 4 BCE–65 CE) even wrote about using "reading stones"—simple magnifying glasses made of polished crystal or glass—to help with reading. However, these were far from the modern glasses we use today.

In the centuries that followed, various cultures, including the Arabs, developed advancements in glassmaking and optics, laying the groundwork for future innovations in vision correction.

The Birth of Eyeglasses: Italy in the 13th Century

The true breakthrough in the development of eyeglasses occurred in Italy, likely in the city of Venice, in the late 13th century. At this time, Venice was a major center of glassmaking, known for its skilled craftsmen and innovative glasswork. Venetian glassmakers were already experimenting with lenses for various uses, including magnifying glasses and burning glasses used for focusing sunlight.

The invention of eyeglasses is often attributed to Salvino D'Armate, a scholar from the city of Padua, who is said to have created the first wearable spectacles around 1286. The story goes that D'Armate developed glasses that could be worn on the face to aid people with nearsightedness (myopia) and farsightedness (hyperopia). These early eyeglasses consisted of two convex lenses held together by a frame that rested on the nose. D'Armate's invention was an immediate success, particularly among monks and scholars, who used them to read and write.

However, it's important to note that the exact details of D'Armate's involvement are debated by historians, and some suggest that the invention of eyeglasses might have been a gradual development involving many different craftspeople and scholars. Nevertheless, Italy is generally considered the birthplace of wearable eyeglasses, with Venice and Padua being key centers for their creation.

Early Design and Development of Eyeglasses

The first eyeglasses were quite rudimentary. The lenses were often made from crystal or quartz, and the frames were simple, sometimes using rivets or a leather band to hold the lenses in place. These early spectacles did not have temples or arms to rest over the ears, so they had to be held up with one hand or placed on the nose and held in place with the other hand.

Over time, Italian glassmakers refined the design and materials of eyeglasses. The lens technology was also improved. By the 14th century, lens grinding techniques were advanced enough that lenses could be made with more precise curvature, improving their ability to correct vision.

The Spread of Eyeglasses

Eyeglasses quickly became popular in Italy and spread throughout Europe. By the 14th century, they were in use among scholars, clergy, and the elite, who used them to read religious texts, legal documents, and academic manuscripts. The popularity of eyeglasses continued to grow, and by the 15th century, they had reached other parts of Europe, including France, England, and Spain.

The development of eyeglasses was an important advancement in the field of optics, and it helped to pave the way for further scientific studies of light and vision. One of the most notable advancements came in the 17th century with the development of the telescope and microscope, both of which relied on advancements in lens-making that had been pioneered in Italy.

Eyeglasses in the Renaissance

During the Renaissance, Italy continued to be the center of eyeglass production. Venice was home to many of the best glassmakers, and Italian craftsmen began to experiment with new materials and designs for eyeglasses. This period saw the introduction of the first temples (arms) on eyeglasses, which helped hold the spectacles securely in place on the ears. These glasses were often made from horn, wood, or metal, and the lenses were ground to higher precision.

Eyeglasses were particularly popular among scholars and intellectuals during this time, and figures like Leonardo da Vinci and Galileo Galilei are believed to have used them. Galileo, in particular, used lenses in his work to make observations of the stars and planets, leading to groundbreaking developments in astronomy.

The story of the invention of the Electric Battery

The story of the first electric battery invention begins with Alessandro Volta, an Italian scientist and inventor, whose groundbreaking work in the late 18th century laid the foundation for the field of electrochemistry and the development of electrical power. Volta's creation of the first electric battery was a milestone in the history of science, and it remains one of the most significant achievements in the study of electricity.

Background: Early Studies of Electricity

Before Volta's time, electricity was a mysterious force that had fascinated scientists and philosophers for centuries. The ancient Greeks had observed the phenomenon of static electricity, and during the 17th and 18th centuries, researchers like William Gilbert and Benjamin Franklin conducted early experiments with electricity, exploring its nature and behavior. However, most of these experiments involved static electricity or electrical discharges, and there was little understanding of how to produce a continuous flow of electricity.

In the 18th century, many scientists were experimenting with electricity and magnetism, but there was no practical way to generate a constant electrical current. At this time, Luigi Galvani, an Italian anatomist and physician, conducted a famous experiment in which he discovered that the muscles of a frog's leg twitched when touched by metal, which he believed was due to an electrical current produced by the metals. This observation led Galvani to propose that animals produced electricity naturally, and he theorized that there was an electrical force in the animal body.

Galvani's findings sparked the interest of many other scientists, but it was Alessandro Volta who would take the next crucial step in understanding electricity and creating the first electrical battery.

Volta's Innovation: The Creation of the Voltaic Pile

In the late 18th century, Alessandro Volta was a professor of physics at the University of Pavia in Italy. Volta was influenced by the work of his predecessor, Luigi Galvani, but he disagreed with Galvani's idea that electricity was produced by animal tissue. Volta believed that the electricity observed in Galvani's experiment was generated by the contact between two different metals, not by the animal's body itself.

In 1800, after years of experimentation, Volta succeeded in creating the first electric battery. His invention, known as the voltaic pile, was a simple but revolutionary device that could produce a steady, continuous flow of electricity.

The voltaic pile was made by stacking alternating discs of two different metals—zinc and copper—separated by thin discs of cardboard or flannel soaked in saltwater (or acid). Each pair of metal discs created a small electrical charge, and when the discs were stacked in a pile, the individual charges added up, creating a continuous flow of electricity.

The voltaic pile was the first practical device that could produce direct current (DC) electricity. By connecting the ends of the pile with wires, Volta demonstrated that the pile could power simple devices, such as a lightbulb, and create sparks similar to those produced by static electricity, but in a continuous and controllable manner.

Volta's invention of the voltaic pile provided the first reliable source of electricity, and it was a pivotal moment in the history of science. Volta's work bridged the gap between early experiments with electricity and the practical applications of electrical power.

The Impact of Volta's Battery

The invention of the voltaic pile was groundbreaking because it demonstrated that electrical energy could be produced chemically, opening up new possibilities for scientific research and practical applications. Volta's battery allowed scientists to study electricity in a more controlled and systematic way, leading to the development of new theories and discoveries in electrochemistry, electromagnetism, and electricity.

Volta's invention also influenced other key developments in the field of electricity:

1. **Chemical Energy and Electrochemistry**: Volta's work showed that chemical reactions could be harnessed to produce electrical energy. This laid the groundwork for later advancements in electrochemical cells and batteries, such as the development of the lead-acid battery and the alkaline battery.

2. **The Development of Electrical Circuits**: Volta's discovery also contributed to the development of electrical circuits, as scientists began to understand how electric current could flow through conductors and power devices.

3. **The Discovery of New Metals and Materials**: Volta's experiments with different metals, including copper and zinc, led to an understanding of how different materials could conduct or resist

electrical current. This knowledge paved the way for the discovery of other essential materials used in modern electronics.

4. **The Creation of the Unit of Voltage**: In recognition of his pioneering work, the unit of electric potential difference, or voltage, was named after Volta. The volt became the standard unit of measurement for electrical potential.

5. **Inspiration for Future Innovators**: Volta's battery inspired other scientists and inventors to continue working on the practical applications of electricity. In the years following Volta's discovery, inventors like Michael Faraday and Thomas Edison would further develop electrical technologies, ultimately leading to the widespread use of electricity in homes and industries.

Volta's Recognition and Legacy

Volta's invention of the first electric battery earned him widespread recognition across Europe and beyond. He was awarded numerous honors and awards, including being named a Count by Napoleon Bonaparte in 1810. Volta continued his work in the field of electricity and made several other important contributions to the study of electrical phenomena.

Today, Volta is remembered as one of the key figures in the history of electricity and physics, and his invention remains one of the most significant scientific breakthroughs of all time. His work not only revolutionized the understanding of electricity but also set the stage for the countless innovations that would follow in the fields of electronics, power generation, and modern technology.

The development of the electric battery was also a major turning point in the industrial revolution, as it provided the means for portable power, which would eventually lead to the development of electric motors, telecommunications, and all the modern applications of electricity that are integral to contemporary life.

***

The invention of the first electric battery by Alessandro Volta in 1800 was a milestone that changed the course of science and technology forever. By creating the voltaic pile, Volta demonstrated that chemical reactions could produce a continuous flow of electrical current, laying the foundation for future advancements in electricity, electronics, and energy storage. Volta's work influenced generations of scientists and engineers, and his legacy continues to be felt today in every battery-powered device and electrical system we use.

The story of the invention of the Maser and Laser technologies

The invention of the maser and laser technologies is a fascinating story that involves a series of breakthroughs in the field of quantum mechanics, electromagnetism, and optics. Although both of these technologies are associated with a few key figures, the involvement of Italians, particularly Renato Elfassi and Giorgio Marconi, played an important role in the evolution of these technologies. Let's dive into the history of how these devices, which would go on to revolutionize fields ranging from medicine to communication, came into being.

The Birth of the Maser and Laser: An Overview

The maser (Microwave Amplification by Stimulated Emission of Radiation) and the laser (Light Amplification by Stimulated Emission of Radiation) are both based on the principle of stimulated emission. This concept was first proposed by Albert Einstein in 1917, when he introduced the idea that an electron in an atom could be "stimulated" to emit radiation. However, it wasn't until much later that practical devices based on this theory were developed.

The maser and laser work by amplifying electromagnetic radiation, either in the microwave (maser) or visible light (laser) spectrum, using a process called stimulated emission. The technology involves excited electrons emitting photons as they return to a lower energy state. When a large number of photons are emitted in phase, they create an intense beam of coherent light or microwaves.

The Invention of the Maser: Italian Contributions

The maser was invented before the laser, and its development is often associated with two scientists, Charles Townes and Arthur Leonard Schawlow, both American physicists. However, the Italian contributions to the development of maser technology are also significant, particularly in microwave amplification.

In the early 1950s, Renato Elfassi, an Italian-American physicist, was conducting research on microwaves and quantum electronics. Elfassi worked closely with Giorgio Marconi, an Italian engineer known for his pioneering work in radio and wireless communication (although Marconi did not directly invent the maser, his earlier work in radio waves laid some of the groundwork for later developments in electromagnetic technologies).

In 1953, Renato Elfassi and Giorgio Marconi independently advanced the understanding of how electromagnetic waves could be amplified using quantum principles, leading to the development of the maser. Their key discovery was the use of ammonia gas in a stimulated emission process to amplify microwave radiation.

Elfassi and Marconi's work in microwave amplification was crucial for the later development of maser devices, which would come into broader use for radar and communications. Maser technology also paved the way for the development of the laser.

The Laser: From Maser to Light

The laser was a natural evolution from the maser, and its development involved both theoretical groundwork and experimental breakthroughs. The laser operates on the same principles as the maser, but it uses light (rather than microwaves) for amplification.

The first working laser was developed by Theodore Maiman in 1960, at Hughes Research Laboratories in California. He created a ruby laser, which used a crystal of ruby (aluminum oxide doped with chromium) as the active medium. While Maiman is credited with creating the first practical laser, Italian scientists made significant contributions to the theoretical aspects and early experimental progress in laser technology.

The concept of the optical laser—light amplification by stimulated emission—was built on the earlier work on masers. Physicist Giovanni Modugno, an Italian physicist who worked in Rome, was among those who contributed to the development of the theoretical framework for laser technology. His work focused on the interaction of atoms with electromagnetic fields, and the early research from his team helped clarify the energy transitions needed to produce a coherent light beam in the laser.

Another notable Italian figure in laser research is Marcello Lippi, who conducted early work in the mid-20th century exploring various types of quantum optics, a field that would later prove vital in laser technology. In fact, Lippi's studies of quantum states of matter and photon interactions helped solidify the theoretical groundwork for laser technology.

The Key Innovations and Italian Influence

The first lasers were based on a variety of materials, including ruby, helium-neon, and semiconductor lasers. These early lasers had diverse applications, including:

- **Scientific research**: Lasers provided a precise way to measure distances and perform high-precision spectroscopy.
- **Telecommunications**: Laser light became the backbone of fiber-optic communication systems, vastly improving the speed and capacity of data transmission.
- **Medicine**: Lasers were used for surgeries, diagnostics, and eye treatments.
- **Manufacturing**: Lasers revolutionized cutting, engraving, and material processing.

In Italy, the advent of laser technologies led to the establishment of laser research institutes and academic programs. The National Research Council of Italy (CNR) became a center for laser research and development, contributing to global advances in both scientific and industrial applications of lasers.

Italian Applications of Laser Technology

Italian scientists and engineers also contributed to the practical applications of laser technologies. For example:

- **Laser-based sensors**: Researchers in Italy played a major role in developing laser-based systems for distance measurements, such as LiDAR (Light Detection and Ranging), which is now used in everything from surveying to autonomous vehicles.

- **Medical lasers**: Italian engineers were pioneers in adapting lasers for use in ophthalmology (eye surgeries) and oncology (treatment of tumors).

- **Laser art and entertainment**: In the field of entertainment, Italian engineers and designers contributed to the creation of laser light shows and projectors used in theaters, theme parks, and public events.

***

Both maser and laser technologies have had profound impacts on numerous fields, including communications, medicine, industry, and scientific research. While Italy did not have a singular inventor of the maser or laser, Italian scientists and engineers played crucial roles in the theoretical development and practical applications of both technologies.

The maser served as the precursor to the laser, and the work done by Italian physicists and engineers—ranging from Renato Elfassi and Giorgio Marconi to Giovanni Modugno and Marcello Lippi—was instrumental in the evolution of these technologies. Through their contributions, they helped usher in the modern age of optics and electromagnetism, shaping the technologies we rely on today.

Day 12: Switzerland

Switzerland, despite its small size, has made remarkable contributions to various fields, especially in engineering, science, medicine, and finance. Here are fifteen important innovations, explorations, and inventions introduced by Swiss people to the world:

1. **Swiss Army Knife**: Invented in the 1890s by Karl Elsener, the Swiss Army knife became a world-renowned multi-tool. It includes various functions like a blade, scissors, screwdriver, and more, revolutionizing personal utility tools worldwide.

2. **Wristwatch**: The wristwatch was popularized and perfected by Swiss watchmakers, including Cartier and Rolex. Swiss watchmaking craftsmanship and innovations, such as the development of automatic and quartz movements, helped establish Switzerland as a leader in precision horology.

3. **Milk Chocolate**: Swiss chocolatier Daniel Peter invented milk chocolate in 1875 by mixing cocoa with condensed milk. His creation transformed chocolate into the sweeter, creamier treat enjoyed worldwide today.

4. **Velcro**: Swiss engineer George de Mestral invented Velcro in 1941 after observing how burrs stuck to his clothing. This hook-and-loop fastening system is used in everything from clothing and shoes to space suits and medical devices.

5. **Direct Democracy**: Switzerland is known for pioneering the practice of direct democracy. Citizens have the right to vote on laws, policies, and constitutional amendments through regular referendums, inspiring similar practices worldwide.

6. **Cellophane**: Cellophane, a transparent packaging material, was invented by Swiss chemist Jacques E. Brandenberger in 1908. This biodegradable material became widely used in food packaging and wrapping.

7. **CERN and Particle Physics**: Switzerland is home to CERN (European Organization for Nuclear Research), the world's leading particle physics research center. The Large Hadron Collider (LHC)

at CERN has contributed significantly to our understanding of particle physics, including the discovery of the Higgs boson.

8. **The Red Cross**: Swiss humanitarian Henry Dunant founded the International Committee of the Red Cross (ICRC) in 1863 after witnessing the horrors of war. The Red Cross set the standard for humanitarian aid and emergency medical assistance.

9. **Dadaism**: Dadaism, an influential avant-garde art movement, was born in Zurich during World War I. Artists like Tristan Tzara and Hugo Ball used absurdity, satire, and collage to critique social norms, inspiring modern and postmodern art.

10. **Helvetica Typeface**: Helvetica, one of the world's most popular typefaces, was designed by Swiss typographers Max Miedinger and Eduard Hoffmann in 1957. Its clean, modern look is widely used in design, advertising, and branding.

11. **Insulin Synthesis**: Swiss chemist Auguste Compagnon contributed to the synthesis and commercialization of insulin for treating diabetes. This development saved millions of lives and transformed diabetic treatment globally.

12. **Nescafé (Instant Coffee)**: Nestlé, a Swiss company, launched Nescafé instant coffee in the 1930s. This innovation made coffee consumption easier and faster, revolutionizing coffee culture worldwide.

13. **The Swiss Banking System**: Switzerland pioneered banking practices such as strict client confidentiality and secure international financial transactions, creating a model for private banking and offshore finance used globally.

14. **The Pax Peace System**: Switzerland has maintained a policy of neutrality and has served as a mediator in international conflicts, promoting global diplomacy and conflict resolution. The Swiss Confederation's approach to neutrality has influenced peacekeeping and diplomatic practices worldwide.

These Swiss contributions reflect a commitment to quality, innovation, and precision. From advancements in food and beverage and medical breakthroughs to the arts, humanitarianism, and finance, Switzerland's legacy of inventions and practices continues to shape our world.

The story of the invention of the Red Cross

The story of the Red Cross's invention is one of humanitarian vision, compassion, and Swiss neutrality. Founded in 1863, the International Red Cross became one of the world's most recognized humanitarian organizations. Its creation was inspired by the desire to provide aid to the wounded and suffering during armed conflicts, regardless of nationality, and to provide relief in times of disaster. The roots of this iconic movement can be traced to Switzerland, with key figures like Henri Dunant playing an essential role in its creation.

The Inspiration: The Battle of Solferino (1859)

The story of the Red Cross begins in 1859, during the Second Italian War of Independence. In the small town of Solferino, located in northern Italy, a fierce battle took place between the Austrian Empire and the combined forces of the Kingdom of Sardinia and France. Over 300,000 soldiers participated in the battle, and the aftermath was devastating. Thousands of men were left wounded on the battlefield, with no medical assistance or supplies to tend to them. There was no system in place to help the wounded, and the casualties were left to suffer or die.

Among the witnesses to the brutality of the battlefield was Henri Dunant, a young Swiss businessman and social activist. Dunant had been traveling in northern Italy on business when he arrived in Solferino just after the battle. What he saw profoundly impacted him: he witnessed thousands of soldiers, both dead and wounded, lying on the battlefield in desperate need of medical attention. Without adequate care, many soldiers were left to die, and the situation was exacerbated by a lack of clean water, shelter, and food.

Dunant was deeply moved by the suffering he witnessed and felt compelled to take action. Along with local villagers, he helped organize makeshift hospitals and coordinated the delivery of food, medicine, and medical supplies to treat the injured. Dunant's compassion for the wounded soldiers was a pivotal moment, as it sparked the idea of a permanent, organized system of humanitarian aid for soldiers on the battlefield.

The Call for Change: "A Memory of Solferino"

After witnessing the horrors of the battle at Solferino, Henri Dunant returned to Switzerland and began to work on his ideas for reform. In 1862, he published a book called "A Memory of Solferino" (originally titled *"Un Souvenir de Solferino"*), in which he described the dire conditions he had seen and outlined a proposal for an international organization to help the wounded in wartime. Dunant suggested the establishment of neutral volunteer groups, trained and prepared to assist soldiers in times of war, regardless of which side they fought for.

Dunant's book also advocated for the establishment of a neutral and impartial organization to provide medical aid and protection to those affected by war, both soldiers and civilians. He proposed that this organization be supported by the major European powers and that it should be governed by a neutral country, ideally Switzerland, to avoid any political conflicts.

His visionary work was widely read and sparked significant discussion across Europe. The idea of a humanitarian organization that could provide impartial aid during war gained momentum. Dunant's work resonated with many reform-minded individuals and leaders in Europe, who saw the potential to provide care to the wounded and ensure that soldiers were treated humanely, regardless of their nationalities.

The Creation of the Red Cross: The Birth of a Movement

In response to Dunant's book and his ideas, a group of Swiss citizens, including Gustave Moynier, William H. Stewart, and Louis Appia, came together in Geneva in 1863 to discuss how they could bring Dunant's vision to life. They founded the International Committee for the Relief of the Wounded Soldiers, which would later become the International Committee of the Red Cross (ICRC). This organization was intended to carry out humanitarian work in times of war, following the principles of impartiality and neutrality.

In 1864, the First Geneva Convention was held, where representatives from several countries gathered to discuss the protection of soldiers and civilians during armed conflicts. The ICRC's proposal to provide neutral medical aid was accepted, and the Geneva Conventions laid the legal foundation for the protection of the wounded and the establishment of the Red Cross as a neutral body. The convention called for the establishment of relief organizations that would provide medical assistance to wounded soldiers on the battlefield, and it provided a set of humanitarian rules to be followed in times of war.

The Symbol of the Red Cross

The Red Cross symbol, which would become synonymous with humanitarian aid, was inspired by the Swiss flag. The flag of Switzerland is a white cross on a red background, and the ICRC adopted this design but reversed the colors. The Red Cross emblem—a red cross on a white background—was chosen to represent the neutral, humanitarian mission of the organization, and it also symbolized the organization's Swiss origins. The emblem was intended to be a symbol of protection, signifying that those wearing the emblem were not to be harmed during conflict.

Expanding the Movement: Global Reach and Influence

The creation of the Red Cross marked the beginning of a new era in international humanitarian law and response. The ICRC quickly expanded beyond Switzerland and attracted support from governments and volunteers worldwide. Over time, the mission of the Red Cross grew to include not only wartime relief but also disaster relief, emergency medical assistance, and other forms of humanitarian aid in times of crisis.

In addition to the International Committee of the Red Cross, national Red Cross societies were established in many countries around the world. The movement soon spread beyond Europe, and by the early 20th century, Red Cross societies existed in most countries. In 1919, the League of Red Cross Societies was established to coordinate efforts between national organizations, further extending the Red Cross's reach.

Today, the International Red Cross and Red Crescent Movement includes the International Federation of Red Cross and Red Crescent Societies (IFRC), which coordinates the humanitarian efforts of national Red Cross and Red Crescent societies across the globe, and the ICRC, which remains headquartered in Geneva, Switzerland.

Henri Dunant: The Founder

Henri Dunant's vision and humanitarian spirit laid the foundation for what would become one of the world's most respected and enduring humanitarian organizations. For his work, Dunant was awarded the first-ever Nobel Peace Prize in 1901, alongside Frédéric Passy, for their efforts to promote peace and humanitarian aid.

The legacy of Henri Dunant and the Red Cross continues to this day. The ICRC operates in conflict zones worldwide, providing medical care, supplies, and protection to those affected by war. The Red Cross movement's principles of neutrality, impartiality, and humanitarian aid have made it an enduring force for good, working tirelessly to alleviate human suffering.

***

The Red Cross is a testament to the Swiss commitment to neutrality, humanitarianism, and international cooperation. From the inspiration of Henri Dunant in Solferino to the founding of the Red Cross in Geneva in 1863, Switzerland's role in the creation of this organization cannot be overstated. The Red Cross continues to serve as a symbol of hope, compassion, and humanity, with Switzerland as its home base, forever connected to the values of human dignity and neutral aid.

Day 13: Germany

Germany has made extensive contributions to science, technology, engineering, philosophy, and the arts. Here are fifteen of the most important innovations, explorations, and inventions introduced by Germans:

1. **Printing Press**: Johannes Gutenberg invented the movable-type printing press around 1440, revolutionizing the spread of information. This innovation enabled mass printing of books, democratizing knowledge and sparking the Reformation and the Enlightenment.

2. **Automobile**: Karl Benz is credited with inventing the first gasoline-powered automobile in 1885. This marked the birth of the automotive industry and transformed transportation worldwide.

3. **X-rays**: Wilhelm Conrad Roentgen discovered X-rays in 1895, earning him the first Nobel Prize in Physics in 1901. This discovery revolutionized medicine by allowing doctors to see inside the human body non-invasively.

4. **Quantum Mechanics**: German physicists, including Werner Heisenberg and Max Planck, were pioneers in developing quantum mechanics, fundamentally changing our understanding of atomic and subatomic particles and influencing physics, chemistry, and technology.

5. **Theory of Relativity**: Albert Einstein, one of Germany's most famous scientists, formulated the theory of relativity, which redefined concepts of space, time, and gravity. His work has had a profound impact on modern physics and cosmology.

6. **Rocket Technology**: Wernher von Braun and his team developed the V-2 rocket during World War II, which became the basis for modern rocketry and space exploration. Von Braun's later work with NASA contributed to the success of the Apollo program.

7. **The Diesel Engine**: Rudolf Diesel invented the diesel engine in 1897. His innovation provided a more efficient alternative to gasoline engines, and diesel engines became essential in industries like shipping, trucking, and construction.

8. **Aspirin**: In 1897, Felix Hoffmann, a chemist at Bayer, synthesized acetylsalicylic acid, which became the active ingredient in aspirin. Aspirin is one of the most widely used medications for pain relief and heart health.

9. **Bacteriology and Microbiology**: Robert Koch made groundbreaking discoveries in microbiology, identifying the pathogens responsible for diseases like tuberculosis, cholera, and anthrax. His work laid the foundation for bacteriology and modern epidemiology.

10. **Fanta**: Created in Germany during World War II due to limited resources, Fanta is a popular soft drink brand developed by Coca-Cola's German branch. Today, Fanta has become a global brand with various flavors.

11. **Kindergarten**: Friedrich Froebel invented the concept of kindergarten in the 19th century, emphasizing play and creativity as essential parts of early childhood education. This concept has been adopted globally and remains a cornerstone of early education.

12. **Modern Organic Chemistry**: Justus von Liebig made significant advancements in chemistry, especially in organic chemistry and agriculture. His work on chemical fertilizers revolutionized farming and increased agricultural productivity worldwide.

13. **The Jet Engine**: Hans von Ohain, a German engineer, was among the inventors of the first operational jet engine, paving the way for modern air travel. This invention led to the development of faster, more efficient aircraft.

14. **MP3 Technology**: Karlheinz Brandenburg, along with his team at the Fraunhofer Institute, developed the MP3 audio compression format, which revolutionized digital audio and the music industry by allowing easy storage and sharing of audio files.

15. **Relational Database Model**: German computer scientist Edgar F. Codd, while working at IBM, created the relational database model. This innovation became the foundation for modern database management systems, impacting everything from banking to online commerce.

These contributions by Germans have left an indelible mark across multiple fields, shaping the modern world in profound ways. From fundamental scientific theories to practical innovations in industry, technology, and education, German ingenuity continues to influence global progress.

The story of the invention of the Automobile

The story of the automobile's invention is a tale of innovation, perseverance, and technical brilliance, with Carl Benz—a German engineer—playing a pivotal role in its creation. His work in the late 19th century led to the development of the first true automobile, which revolutionized transportation and laid the foundation for the modern automotive industry.

The Early Beginnings: The Quest for a New Form of Transportation

The idea of creating a self-propelled vehicle had been around for centuries. Early inventors such as Nicolas-Joseph Cugnot in France and Richard Trevithick in England built experimental steam-powered vehicles in the 18th and early 19th centuries, but these early machines were slow, cumbersome, and unreliable. Their lack of practicality and efficiency meant that they never gained widespread use.

By the mid-19th century, there was a growing need for a more efficient and reliable form of personal transportation. As industrialization progressed, cities were expanding, and the limitations of horse-drawn carriages were becoming evident. Engineers and inventors around the world began experimenting with different ways to make faster, more practical modes of transport.

Carl Benz: The Innovator Behind the Automobile

Carl Benz was born in 1844 in Germany, and he displayed an early aptitude for mechanics. After studying mechanical engineering at the Karlsruhe Polytechnical School, he worked in various engineering firms and later co-founded his own company, Benz & Cie., in 1883. Benz was not just an inventor but also a businessman, and he was determined to create a practical and efficient vehicle that could replace horse-drawn carriages.

The Invention of the Benz Patent-Motorwagen

Carl Benz's breakthrough came in 1885, when he built the first version of what would become known as the Benz Patent-Motorwagen. This vehicle was powered by an internal combustion engine, a relatively new technology at the time. Benz's design was a major departure from previous efforts, as he focused on using a gasoline-powered engine rather than steam. This made the vehicle lighter, more efficient, and more practical for everyday use.

The Benz Patent-Motorwagen, which he completed in 1885, was a three-wheeled vehicle that featured a single-cylinder four-stroke engine, capable of producing just 0.75 horsepower. It was the first true automobile powered by an internal combustion engine and could reach a top speed of about 10 miles per hour (16 km/h). The engine was mounted at the rear of the vehicle, and it was designed with a simple frame made of steel and wood.

In 1886, Benz was granted a patent for his invention, officially marking the birth of the automobile. The Patent No. 37435 for the "vehicle with gas engine" was granted by the German government, making it the first officially recognized automobile in history.

The First Public Test Drive: Bertha Benz's Historic Journey

In 1888, a pivotal moment in the history of the automobile occurred when Bertha Benz, Carl Benz's wife, decided to take the Benz Patent-Motorwagen on the first long-distance road trip. This journey was not just a test of the vehicle's capabilities but also a bold statement about the future of personal transportation.

Bertha Benz set out from Mannheim to Pforzheim, a distance of about 66 miles (106 kilometers), to visit her mother. This journey was significant because, at the time, no one knew whether the vehicle would be able to handle the distance or even make it through the terrain. There were no gas stations or proper roads for automobiles, and Bertha's trip was made without the knowledge or approval of her husband, Carl.

During the journey, Bertha encountered several challenges, including mechanical failures and obstacles such as hills and narrow roads. However, she was resourceful, making repairs along the way, including using a hairpin to unclog a fuel line and even stopping at a local pharmacy to buy fuel. Bertha's journey was a success and proved the practicality of Benz's invention. It garnered significant media attention and led to increased public interest in the automobile.

The Rise of the Automobile Industry

After Bertha's historic trip, Carl Benz's invention began to gain recognition, and he began to sell the Benz Patent-Motorwagen to the public. In 1889, the Benz & Cie. company was officially formed, and it started producing automobiles on a larger scale. The company's success grew rapidly as it gained customers not only in Germany but also in other parts of Europe.

In 1893, Benz introduced a four-wheeled version of his vehicle, which had greater stability and was more suitable for mass production. By the end of the 19th century, Benz's cars were in high demand, and the automobile industry was beginning to take shape.

The Impact of Carl Benz's Invention

Carl Benz's work paved the way for the global automobile industry, but his invention was just the beginning. The success of the Benz Patent-Motorwagen inspired other inventors and engineers to build upon Benz's design. In particular, the German engineer Gottlieb Daimler and his partner Wilhelm Maybach made key improvements to the automobile, such as introducing the high-speed gasoline engine.

In 1926, Benz & Cie. and Daimler-Motoren-Gesellschaft (the company founded by Daimler) merged to form Daimler-Benz, which would later become the well-known Mercedes-Benz brand.

Legacy of the Automobile

Carl Benz's invention of the automobile is regarded as one of the most important developments in human history. It revolutionized the way people lived, worked, and traveled. The automobile became an essential part of modern society, changing the face of cities, economies, and cultures across the globe.

Today, Mercedes-Benz is a leading global automobile manufacturer, and Benz's contribution to the automobile industry is celebrated around the world. Benz is often referred to as the "father of the automobile", and his work remains a symbol of German ingenuity, engineering excellence, and innovation.

***

Carl Benz's invention of the automobile marked a turning point in history. From the first Patent-Motorwagen in 1886 to the global dominance of the automobile industry, Benz's invention shaped the modern world in profound ways. His legacy lives on not just in the cars we drive today, but also in the engineering principles and innovations that continue to drive the industry forward. Thanks to his vision and determination, Carl Benz's name is forever linked to the birth of the automobile and the evolution of transportation.

The story of the discovery of X-ray

The story of X-ray exploration is one of groundbreaking scientific discovery, and it is deeply linked to Germany, specifically to the work of Wilhelm Conrad Roentgen, a German physicist. Roentgen's invention of the X-ray in 1895 marked a pivotal moment in medical imaging and changed the world of medicine, science, and technology.

The Discovery of X-rays

In 1895, Wilhelm Roentgen was conducting experiments with cathode rays, a type of electron stream, at the University of Würzburg in Germany. He was working with a vacuum tube, which was an apparatus used to study cathode rays. During these experiments, Roentgen noticed something unusual: a fluorescent screen in his lab began to glow, even though it was not in the direct path of the cathode rays. Intrigued by this, Roentgen began to investigate the phenomenon further.

He found that the glowing screen was being affected by a type of radiation that had not been seen before. This radiation, which he could not explain at the time, was able to pass through solid objects like paper, wood, and even human tissue, but it could not pass through denser materials like bone or metal. Roentgen initially called this unknown radiation "X-rays," using the mathematical term "X" to represent something unknown.

Roentgen quickly realized the significance of his discovery. X-rays had the ability to produce images of the interior of solid objects without the need for surgery or dissection. Roentgen's experiments were a breakthrough, as no one had previously developed a method to view the inside of the human body without cutting it open. He saw the potential for X-rays to be used in medical diagnostics, although he did not immediately fully grasp the scope of their potential applications.

The First X-ray Image

Roentgen's first X-ray image, taken in 1895, was of his wife, Anna Bertha Roentgen's hand. The image clearly showed her bones and wedding ring, and this photograph is considered to be the first medical X-ray image ever taken. It was a remarkable achievement, as it demonstrated the ability of X-rays to penetrate soft tissue and reveal the structure of bones inside the body.

This first X-ray image of the hand shocked the scientific community, and the news of Roentgen's discovery spread rapidly. Scientists and doctors across the world recognized the enormous potential of X-rays for medical diagnostics, and this discovery would go on to revolutionize medicine.

Roentgen's Recognition and the Spread of X-rays

In 1896, just a year after Roentgen's discovery, the use of X-rays began to spread around the world. Scientists and doctors began experimenting with X-ray machines to visualize broken bones, locate foreign objects inside the body, and identify diseases. Roentgen's discovery was transformative because it provided a non-invasive method for seeing inside the human body.

Roentgen's discovery was also met with widespread recognition. In 1901, just a few years after his groundbreaking work, Roentgen was awarded the first-ever Nobel Prize in Physics for his discovery of X-rays. His work was considered one of the most important scientific breakthroughs of the 19th century, and it would continue to have far-reaching impacts in the years to come.

X-ray Technology and Medical Advancements

The development of X-ray technology continued rapidly after Roentgen's initial discovery. Early X-ray machines were large, cumbersome, and dangerous due to their high radiation levels. However, over time, scientists and engineers developed more sophisticated and safer X-ray machines.

By the early 20th century, X-rays were being used routinely in medical diagnostics. The ability to view internal organs and bones without surgery was a game-changer in medicine. X-rays allowed doctors to diagnose fractures, tumors, and foreign objects inside the body with remarkable precision. Over the years, medical imaging continued to improve, leading to the development of advanced techniques like CT scans (computed tomography), MRI (magnetic resonance imaging), and ultrasound.

***

Today, X-rays remain an essential tool in modern medicine, used for diagnostic imaging in hospitals and clinics around the world. Roentgen's discovery not only transformed medicine but also had significant implications for fields like physics, materials science, and even art. X-ray technology is used in a wide variety of industries, from detecting flaws in materials to studying ancient artifacts.

Roentgen's Nobel Prize in 1901 and his lasting legacy in science and medicine have ensured that his name is forever linked to the discovery of X-rays. His contribution continues to save countless lives by allowing doctors to diagnose and treat patients with greater accuracy.

The discovery of X-rays by Wilhelm Conrad Roentgen in 1895 was one of the most significant milestones in the history of science. It changed the landscape of medicine, providing a revolutionary new tool for seeing inside the human body. Roentgen's curiosity, experimentation, and dedication to his work led to one of the most transformative discoveries in modern science, and his invention continues to shape medical practices and technologies today.

In recognition of his work, Wilhelm Roentgen is remembered as one of the most important scientists in the world, and his discovery of X-rays remains a cornerstone in both the scientific and medical communities.

The story of the invention of the Jet engine

The story of the jet engine invention is a remarkable tale of engineering ingenuity, and it is deeply tied to Germany, with two key figures—Hans von Ohain and Frank Whittle—who independently developed the concept of the jet engine. However, the first practical jet engines were developed in Germany during World War II, and their creation marked a major technological leap in aviation.

Early Developments in Jet Propulsion

The idea of jet propulsion—using the force of expelled gases to drive a vehicle forward—has ancient roots. Concepts of jet engines date back to the 19th century, with Sir Isaac Newton's Third Law of Motion (every action has an equal and opposite reaction) providing the theoretical foundation for modern jet propulsion.

However, it wasn't until the 20th century that the technology began to take shape. Throughout the early 1900s, a variety of aviation pioneers worked on various forms of aircraft propulsion, from piston engines to rocket motors. The critical development of the jet engine began to take place in the 1930s and 1940s, a time of intense competition between nations, especially during the years leading up to and during World War II.

The German Contribution: Hans von Ohain

While Frank Whittle, a British engineer, is often credited with the invention of the jet engine in terms of theoretical design and patents, Hans von Ohain, a German engineer, is often regarded as the first person to successfully build and fly a jet-powered aircraft.

In the early 1930s, von Ohain was working on an idea to create a more efficient and powerful form of propulsion for aircraft. At that time, aircraft engines were typically powered by piston engines, which were limited in terms of speed and efficiency, especially at high altitudes. Von Ohain believed that a turbine engine—which expelled exhaust gases to create thrust—could offer greater performance.

In 1936, von Ohain joined Heinkel Flugzeugwerke, a German aircraft manufacturer, to further develop his jet engine design. He built a prototype of his engine, the HeS 1, which was a turbojet. His design was a complete departure from the piston engines of the time and focused on harnessing the power of a turbine-driven compressor to generate thrust.

By 1939, von Ohain's engine was successfully installed in the Heinkel He 178, the world's first jet-powered aircraft. On August 27, 1939, the Heinkel He 178 made its maiden flight with von Ohain as the pilot. This marked the first time in history that an aircraft was powered by a jet engine, and it represented a pivotal moment in aviation technology.

Frank Whittle: The British Contribution

At almost the same time, in Britain, Frank Whittle, an engineer and pilot, was also independently working on the concept of a jet engine. Whittle's focus was on developing a turbojet engine that would be more efficient than conventional piston engines. He patented his design in 1930 and received government backing for the development of his ideas in the mid-1930s.

In 1937, Whittle's company, Power Jets Ltd., began to build a prototype engine based on his design. The first successful test of his jet engine took place in 1939, and in 1941, the Gloster E.28/39 became the first British jet-powered aircraft to fly.

While Whittle's engine was technically the same basic concept as von Ohain's, it was Whittle's work in England that led to the development of the first operational jet engine for military aircraft, which became crucial during World War II.

The First Operational Jet Aircraft: The Messerschmitt Me 262

By the time World War II had erupted, Germany's aircraft engineers had made significant strides in jet propulsion. Hans von Ohain's engine design was further developed and improved, and by 1943, it was used in the Messerschmitt Me 262, the world's first operational jet fighter.

The Me 262 was powered by two Jumo 004 turbojet engines, developed by Junkers, a German company. The aircraft made its first flight in 1942, and its combat debut came in 1944. The Me 262 was faster, more maneuverable, and had a higher altitude capability than anything the Allies had at the time, making it a formidable weapon. However, its impact on the war was limited because of production delays and the challenges of the wartime German industrial infrastructure.

Despite its advanced technology, the Me 262 was introduced too late in the war to turn the tide of battle. Nevertheless, it demonstrated the immense potential of jet-powered aircraft, and its development laid the foundation for post-war military aviation.

The Post-War Influence and Jet Age

After the end of World War II, Germany's advanced jet engine technology was seized by both the Allied powers and the Soviet Union as part of reparations and military spoils. Both sides began working on further developing and refining the jet propulsion technology for use in their own military and civilian aircraft.

In the United States, for example, the Allied occupation forces took the German jet aircraft designs, including the Me 262 and the Heinkel He 162, and began developing their own jet-powered aircraft, such as the North American F-86 Sabre and the McDonnell Douglas F-4 Phantom.

In the years following the war, jet propulsion became the foundation for the jet age in civilian aviation, revolutionizing air travel by dramatically increasing the speed, efficiency, and safety of commercial airplanes.

***

Today, the jet engine is a critical part of modern aviation, powering everything from military fighter jets to commercial airliners. The development of the jet engine has led to faster, more efficient, and safer air travel and is a testament to the brilliant minds like Hans von Ohain and Frank Whittle.

While von Ohain's success in building and flying the first jet-powered aircraft is often overshadowed by Whittle's theoretical contributions, both men played essential roles in the development of jet propulsion technology. Germany's role, particularly through the work of von Ohain, is central to the history of the jet engine, and it remains one of the most significant contributions to modern aviation. The Messerschmitt Me 262, powered by the Jumo 004 engine, is still recognized as a historic breakthrough in the field of aerospace engineering, and its development during World War II helped usher in the era of jet-powered flight.

Day 14: Denmark

Denmark has contributed significantly to global progress through innovations, explorations, and inventions across various fields, including science, technology, medicine, and design. Here are fifteen of the most impactful contributions introduced by the Danish:

1. **Insulin Production**: Danish scientists were instrumental in the development and commercialization of insulin. In the 1920s, Nobel Laureate August Krogh, with support from H.C. Hagedorn, set up a facility in Denmark to produce insulin, making it widely accessible to treat diabetes.

2. **LEGO**: One of Denmark's most famous contributions, LEGO bricks were invented by Ole Kirk Christiansen in 1932. The interlocking plastic bricks have become a global cultural phenomenon, encouraging creativity and innovation in play for generations of children and adults.

3. **pH Scale**: Danish chemist Søren Sørensen introduced the pH scale in 1909 while working at the Carlsberg Laboratory. This measurement system, which gauges the acidity or alkalinity of substances, is now fundamental in chemistry, biology, and environmental science.

4. **Wind Turbines**: Denmark has been a pioneer in wind energy technology. In the 1970s, Danish engineers developed modern wind turbines, which helped Denmark become a global leader in renewable energy and set standards in green technology worldwide.

5. **Magnetic Tape Recording**: Danish engineer Valdemar Poulsen invented the first magnetic sound recorder, the telegraphone, in 1898. This innovation laid the groundwork for modern magnetic tape recording, revolutionizing the music, broadcast, and data storage industries.

6. **Telegraphone (First Sound Recorder)**: Valdemar Poulsen's invention, the telegraphone, was the first device to record and play back sound magnetically. This breakthrough technology

influenced the development of magnetic recording and data storage technologies.

7. **High-Efficiency Insulin**: Novo Nordisk, a Danish pharmaceutical company, has been at the forefront of insulin and diabetes care innovations, developing high-efficiency insulin products that continue to advance diabetes treatment worldwide.

8. **Grundtvigian Education Model**: N.F.S. Grundtvig, a Danish philosopher, theologian, and educational reformer, advocated for a unique "folk high school" system focused on non-formal, life-oriented education. This model influenced adult education globally, especially in Scandinavian countries.

9. **Telemedicine**: Denmark has been a leader in telemedicine and electronic health records, especially since the 1990s. The Danish healthcare system uses advanced digital platforms to provide remote healthcare services, a model many countries are now adopting.

10. **Lego Robotics**: The LEGO Mindstorms robotics kits, developed in collaboration with the Massachusetts Institute of Technology (MIT), introduced programmable robotics to young students and hobbyists, encouraging STEM education worldwide.

11. **Sea Navigation and Exploration**: Danish explorers like Vitus Bering led major expeditions to map new territories, including parts of Russia and the Arctic. Bering's exploration paved the way for trade routes and expanded knowledge of the North Pacific and Arctic regions.

12. **Bluetooth Technology**: Named after the Viking king Harald Bluetooth, who unified Denmark and Norway, Bluetooth technology was developed by Danish engineers working with Ericsson. This wireless communication technology has become essential in connecting devices globally.

13. **Hand Hygiene and Public Health Initiatives**: Inspired by Danish reformer F.F. Ulrik Langen's public health initiatives, Denmark prioritized hand hygiene and disease prevention early in the 20th century, setting standards that influenced hygiene practices worldwide.

From renewable energy to modern design, telecommunications, and medicine, Danish innovations reflect a commitment to practicality, sustainability, and social responsibility. These contributions continue to influence industries, improve public health, and shape culture on a global scale.

The story of the invention of the Bluetooth

The story of Bluetooth and its invention is closely tied to Denmark, where a Danish engineer, Jaap Haartsen, and his team at Ericsson, a telecommunications company, pioneered the technology that would later become ubiquitous in wireless communication devices.

The Roots of Bluetooth Technology

In the 1990s, the world was undergoing rapid advancements in wireless communication, but there was still a lack of a universal, short-range communication standard that could allow devices to connect wirelessly over short distances. At the time, there were several proprietary wireless technologies, but there was no universally accepted standard that could seamlessly link devices like mobile phones, headsets, computers, and personal digital assistants (PDAs).

Ericsson, a Swedish telecommunications company with a large presence in Denmark, was particularly interested in finding a solution that would enable communication between devices without the need for physical cables. Jaap Haartsen, a Dutch engineer working at Ericsson's Research and Development center in Lund, Sweden, became a key figure in this process.

Jaap Haartsen and the Development of Bluetooth

In 1994, Haartsen and his colleague Sven Mattisson at Ericsson began working on the development of a new short-range radio technology. They wanted to create a way for electronic devices to communicate over short distances—up to about 100 meters—without the need for wires. Their goal was to develop a standard that would be low-cost, low-power, and capable of handling data and voice communications across a wide range of devices.

Haartsen's team developed a radio-frequency technology that could transmit data between devices over short distances, which was based on a 2.45 GHz frequency band. This band was already widely available for short-range devices like microwave ovens and cordless phones, which made it ideal for Bluetooth's low-power communication requirements.

The name Bluetooth itself is inspired by a historical reference: King Harald "Bluetooth" Gormsson, a 10th-century Viking king who united Denmark and Norway. Haartsen and his colleagues chose this name because they saw the technology as a way to unite different devices and make them communicate seamlessly. The Bluetooth logo itself incorporates a combination of the initials of King Harald "Bluetooth" (H and B in the Nordic runic alphabet).

Bluetooth's Initial Release and Expansion

In 1998, after several years of development, the Bluetooth Special Interest Group (SIG) was formed by several major companies, including Ericsson, Intel, Nokia, and IBM, to oversee the further development and adoption of Bluetooth technology. The SIG's formation marked a key milestone in the technology's journey, as it provided a unified effort to standardize Bluetooth and promote its use across various industries.

The Bluetooth standard was officially introduced in 1999, and by the early 2000s, the first Bluetooth-enabled devices began to appear on the market. These early devices included Bluetooth headsets, mobile phones, and laptops, which could wirelessly communicate with one another within short ranges. Bluetooth technology quickly gained traction for its convenience and versatility, allowing people to wirelessly connect and interact with their devices without the hassle of tangled cables.

The Impact and Growth of Bluetooth

Bluetooth technology soon became indispensable for connecting a wide variety of devices. Its low power consumption and versatility made it ideal for a variety of applications, from wireless audio and hands-free calling to file transfers and gaming. Bluetooth's standardization meant that it could be implemented in a wide range of consumer electronics, regardless of the brand.

Bluetooth also evolved with each new generation, increasing its range, speed, and efficiency. Over time, Bluetooth became an essential part of wireless communication, appearing in devices such as smartphones, wireless speakers, fitness trackers, smartwatches, and even home appliances.

By the mid-2000s, Bluetooth had become a critical component of modern technology, and it is now used in billions of devices worldwide. Bluetooth technology has not only reshaped personal communication but also enabled the growing field of the Internet of Things (IoT), where everyday devices are interconnected and can communicate with one another wirelessly.

***

Jaap Haartsen's work on Bluetooth technology has left a lasting legacy. He is often credited with the invention of Bluetooth, and his contributions were recognized in 2012 when he received the prestigious IEEE Masaru Ibuka Consumer Electronics Award for his role in the development of Bluetooth. Haartsen's invention has transformed the way we use electronic devices, making wireless communication ubiquitous and improving our everyday lives.

Today, Bluetooth is a cornerstone of modern technology, and it continues to evolve, with the most recent versions offering even faster speeds, longer ranges, and support for new applications like wireless audio streaming and connectivity for smart homes and cities.

The invention of Bluetooth in Denmark by Jaap Haartsen and his colleagues at Ericsson is one of the most influential technological breakthroughs of the late 20th century. It revolutionized the way devices connect and communicate, paving the way for the wireless world we live in today. From mobile phones and computers to home automation systems and medical devices, Bluetooth has become an essential part of modern life.

Bluetooth's journey from a small, experimental idea to a global communication standard highlights the importance of collaboration across borders and industries. The legacy of Bluetooth technology continues to shape our digital world, and it all began with the vision of Jaap Haartsen in Denmark in the 1990s.

Day 15: Sweden

Sweden has contributed extensively to fields such as technology, medicine, science, and culture. Here are fifteen of the most impactful innovations, explorations, and inventions introduced by Swedes to the world:

1. **Dynamite**: Alfred Nobel, a Swedish chemist and engineer, invented dynamite in 1867, revolutionizing construction and mining. Nobel's fortune from this invention led to the establishment of the Nobel Prizes, one of the world's most prestigious awards for achievements in science, literature, and peace.

2. **The Nobel Prize**: Established by Alfred Nobel in his will, the Nobel Prizes honor outstanding contributions in physics, chemistry, medicine, literature, and peace. Since 1901, the Nobel Prizes have celebrated global advancements in various fields, encouraging innovation and humanitarian efforts.

3. **Three-Point Seatbelt**: Swedish engineer Nils Bohlin invented the three-point seatbelt in 1959 while working at Volvo. This safety device has saved millions of lives and is a standard feature in modern automobiles worldwide.

4. **Tetra Pak**: In 1946, Ruben Rausing developed Tetra Pak, a revolutionary packaging solution that allows liquid products like milk and juice to be stored safely without refrigeration. Tetra Pak has transformed food distribution and storage globally.

5. **Pacemaker**: Swedish engineer Rune Elmqvist invented the first implantable pacemaker in 1958. This device has since saved countless lives by helping patients with heart conditions regulate their heartbeat.

6. **The Celsius Temperature Scale**: Anders Celsius, a Swedish astronomer, developed the Celsius temperature scale in 1742. This temperature scale is used worldwide in scientific contexts and everyday life, particularly outside the United States.

7. **Skype**: Originally developed by Swedes Niklas Zennström and Dane Janus Friis in 2003, Skype transformed communication by

making video calls widely accessible, paving the way for modern video conferencing platforms.

8. **Ultrasound in Medicine**: Swedish physician Inge Edler and engineer Carl Hellmuth Hertz pioneered the use of ultrasound in medicine in the 1950s. Ultrasound technology is now essential in medical diagnostics, especially in obstetrics and cardiology.

9. **Adjustable Wrench**: The adjustable wrench, commonly known as a "Swedish key" or "Swedish wrench," was invented by Johan Petter Johansson in 1892. This versatile tool is used worldwide in mechanical and construction work.

10. **IKEA**: Founded by Ingvar Kamprad in 1943, IKEA has popularized affordable, flat-pack furniture that can be assembled at home. IKEA's minimalist Scandinavian designs have transformed the furniture industry globally.

11. **Hasselblad Cameras**: Victor Hasselblad designed high-quality cameras that gained fame after being used by NASA during the Apollo moon missions. Hasselblad cameras are renowned for their precision and durability in photography.

12. **Ball Bearing**: Swedish engineer Sven Wingquist invented the modern self-aligning ball bearing in 1907. This innovation revolutionized machinery and enabled smoother, longer-lasting mechanical movements in countless applications.

13. **The Propeller**: John Ericsson, a Swedish engineer, invented the ship propeller, which improved ship speed and efficiency. His design was fundamental in the development of modern marine engineering.

14. **Spotify**: Founded by Daniel Ek and Martin Lorentzon in 2006, Spotify revolutionized music streaming, providing users access to a vast library of songs and enabling artists to reach global audiences.

15. **Zipper**: The modern zipper was refined and made commercially viable by Swedish-American engineer Gideon Sundback in 1913. The zipper has become one of the most widely used fastening devices, integral to fashion, luggage, and countless other applications.

These Swedish contributions reflect a legacy of ingenuity in fields as diverse as safety, medical technology, communication, and design. Swedish innovations continue to improve daily life, drive technological progress, and influence global industries.

The story of the invention of the Pacemaker

The story of the invention of the pacemaker is a remarkable tale of medical innovation that began in Sweden in the 1950s. The pacemaker, a device that helps regulate the heart's rhythm, has saved countless lives and revolutionized the treatment of heart conditions. The key figure behind the invention of the first implantable pacemaker was Anders A. Lundqvist, a Swedish engineer, although the story also involves the pioneering work of Dr. Paul Zoll, an American cardiologist, and the work of Rune Elmqvist, a Swedish inventor.

The Early Developments in Heart Rhythm Management

Before the pacemaker, there were limited options for treating people with heart rhythm problems, specifically bradycardia (abnormally slow heart rate), which can lead to symptoms like dizziness, fainting, and in extreme cases, sudden cardiac arrest. Artificial stimulation of the heart's electrical activity had been explored in the early 20th century, but it wasn't until the mid-1900s that significant breakthroughs were made.

Dr. Paul Zoll and External Defibrillation

One of the earliest pioneers in heart rhythm management was Dr. Paul Zoll, an American cardiologist. In the late 1940s, Dr. Zoll developed an external pacemaker that could apply electrical impulses to a patient's chest to regulate their heart rate. This external device was used in emergencies to help patients whose hearts had stopped or whose heartbeats were irregular.

Though it was a significant step forward, the external pacemaker had limitations because it required the patient to be connected to the machine at all times, which was not a practical solution for long-term treatment. The idea of creating an implantable pacemaker, a device that could be placed inside the body to regulate heartbeats continuously, was the next major challenge.

The First Implantable Pacemaker: Rune Elmqvist's Invention

The breakthrough came in Sweden, where Rune Elmqvist, a Swedish engineer, is credited with inventing the first implantable pacemaker in 1958. Elmqvist was working in collaboration with Dr. Åke Senning, a Swedish cardiac surgeon, at Karolinska University Hospital in Stockholm.

Dr. Senning had been treating patients with heart rhythm problems, but the existing methods, like external pacing, were not effective for long-term care. Senning and Elmqvist decided to work together to create a pacemaker that could be implanted inside the body, allowing patients to live with a constant, internal device that could regulate their heart rate.

In 1958, Elmqvist succeeded in creating the first implantable pacemaker. It was a large device compared to modern pacemakers, and it was powered by external batteries. This device was successfully implanted in a patient in Stockholm by Dr. Senning, marking a landmark moment in medical history. It provided the heart with electrical impulses to regulate its rhythm, and the patient survived for several years with the device.

Advancements and Miniaturization

While the first pacemaker was a major milestone, the technology needed further refinement. Over the next few decades, pacemakers became progressively smaller, more efficient, and more reliable. The early models required a larger power source, but the miniaturization of the components, particularly the battery, played a critical role in the development of implantable pacemakers.

In the 1960s, Anders A. Lundqvist, a Swedish engineer working at Crown Medical, made significant contributions to the pacemaker's design. He developed smaller, more reliable batteries and improved the device's function, making it feasible for long-term use inside the body.

The First Long-Term Implantation

In 1960, the first successful implantation of a long-term pacemaker occurred. A patient named Arne Larsson became the first person to live with a pacemaker for an extended period. Dr. Senning performed the procedure, and the patient went on to live for another 43 years, undergoing several pacemaker replacements along the way. Arne Larsson's case is often regarded as the first long-term success of an implanted pacemaker, demonstrating the potential of the device to improve and prolong life.

The Role of Modern Pacemakers

Since the 1960s, pacemakers have evolved tremendously, with modern devices now being implantable, miniature, and powered by long-lasting batteries. These devices can be programmed to monitor and regulate the heart's rhythm, and they have advanced features such as the ability to synchronize the heart's chambers, adjust pacing depending on activity level, and even monitor a patient's heart health remotely.

Today, pacemakers are used not only to treat bradycardia but also in patients with heart failure and arrhythmias. The global adoption of pacemaker technology has significantly improved the quality of life for millions of people worldwide.

Conclusion: The Legacy of the Pacemaker

The invention of the pacemaker is a testament to the power of collaboration between engineers and medical professionals. Rune Elmqvist and Dr. Åke Senning's work in Sweden laid the foundation for the modern pacemaker, and their pioneering efforts have saved countless lives since. The ongoing advancements in pacemaker technology continue to make heart care more effective and personalized.

Today, pacemakers are small, sophisticated devices that provide life-saving support for patients with heart rhythm disorders, and their development remains one of the most significant achievements in the field of cardiac medicine. The Swedish contributions to pacemaker technology have had a profound and lasting impact on the world of medicine, improving the lives of millions.

Day 16: Norway

Norway has a rich history of pioneering explorations, inventions, and innovations that have had a lasting impact across various fields, including maritime exploration, technology, and environmental sciences. Here are fifteen of the most notable contributions introduced by Norwegians to the world:

1. **Telemark Skiing Technique**: Norwegian Sondre Norheim invented the Telemark skiing technique in the 1860s. This skiing style laid the foundation for modern skiing techniques and helped make skiing a popular recreational and competitive sport worldwide.

2. **Paper Clip**: Johan Vaaler, a Norwegian inventor, is credited with designing one of the first versions of the paper clip in the late 19th century. This small but practical tool remains essential for organizing papers in offices and schools.

3. **Gas Turbine**: Norwegian inventor Ægidius Elling developed one of the first functional gas turbines in 1903, which contributed to the development of gas turbine technology now used in aircraft engines and power generation.

4. **Aquavit (Akvavit)**: Aquavit, a traditional Scandinavian distilled spirit, was popularized in Norway. Known for its caraway flavor, aquavit has become an essential part of Nordic culture and culinary heritage.

5. **Fish Farming (Aquaculture)**: Norway is a leader in aquaculture, especially in farming Atlantic salmon. Norwegian innovations in fish farming have made it possible to produce seafood sustainably, feeding millions and influencing aquaculture practices globally.

6. **Svalbard Global Seed Vault**: Located on a Norwegian island, the Svalbard Global Seed Vault stores a vast variety of plant seeds from around the world. It serves as a backup to protect the genetic diversity of crops, ensuring food security in case of global crises.

These innovations and explorations from Norway showcase the country's contributions to science, technology, sustainability, and global well-being. Whether through advancements in healthcare, sustainable practices, or groundbreaking explorations, Norwegian ingenuity continues to leave a lasting legacy on the world.

The story of the invention of the Gas turbine

The invention of the gas turbine is closely associated with the contributions of Norwegian engineer Aga Eide, but it is also deeply rooted in a broader European effort in the early 20th century to develop more efficient and powerful engines, particularly for aircraft and industrial applications.

Early Developments: The Rise of Turbine Technology

The gas turbine is an engine that converts chemical energy (usually from fuel) into mechanical energy using a rotating turbine. Unlike the steam turbine, which was invented earlier, the gas turbine operates using compressed air and burning fuel to create high-pressure exhaust gases that spin a turbine, which in turn drives a generator or a mechanical load.

While many countries were experimenting with turbine technology in the early 20th century, Norway played a significant role in its development, particularly in the context of Aga Eide's contributions.

The Pioneering Work of Aga Eide

In 1910, Aga Eide, a Norwegian engineer, is often credited with developing the first operational gas turbine. Eide's design was based on the concepts of expanding high-pressure gases that were already in use in steam turbines. However, Eide's key innovation was utilizing combustion inside the turbine itself, which made it more efficient than steam turbines for certain applications.

The technology that Eide developed, however, was not immediately ready for mass production. He faced many challenges in perfecting the combustion process and controlling the high temperatures and pressures involved in the turbine's operation. But his work laid the groundwork for future advancements in turbine engines.

Further Developments: Jet Engines and Industrial Power

The evolution of the gas turbine continued throughout the 1930s and 1940s. While Aga Eide made an important early contribution to the concept, it was the work of Sir Frank Whittle (UK) and Hans von Ohain (Germany) in the development of jet propulsion engines during the Second World War that took the technology to the next level. These two engineers developed the first jet engines, which used gas turbines to power aircraft.

While gas turbines were being used for jet propulsion and aircraft engines, there was also a growing interest in adapting them for industrial use. In Norway, engineers began applying the technology to power large generators and machinery in oil platforms, ships, and power plants, further demonstrating the versatility of gas turbines beyond aviation.

Norwegian Contributions to Gas Turbine Technology

Norwegian industry, particularly oil and gas, adopted gas turbines to power offshore drilling platforms and other industrial equipment. Companies like Aker Solutions and Kongsberg Gruppen played an important role in integrating gas turbine technology into offshore and marine industries, making them crucial for Norway's energy sector. Norway's large reserves of oil and natural gas made it one of the leading countries in the use of gas turbines in energy production.

In the latter half of the 20th century, Norway became a global leader in using gas turbines for energy generation. The country invested heavily in offshore oil extraction, using advanced turbine technology to provide power to remote oil rigs. By the 1980s and 1990s, gas turbines had become indispensable to the country's energy infrastructure, particularly for its hydroelectric plants and oil platforms.

Modern Gas Turbines

Today, gas turbines are widely used around the world, not just for powering aircraft but also for electricity generation, industrial machinery, and marine applications. Modern gas turbines are highly efficient, can be powered by various types of fuel, and are key in addressing global energy demands. They can be found in power plants, factories, ships, and even in some electric vehicles.

While Aga Eide's work in the early 20th century was foundational in the development of the gas turbine, the technology was shaped and advanced by engineers around the world. In Norway, gas turbines continue to play a significant role in both energy production and the oil & gas industry, making it a crucial part of the country's economy.

***

The development of the gas turbine is a multi-national story, with Norwegian engineer Aga Eide playing an important role in the early stages. Today, gas turbines power many aspects of modern industry, and Norway, as a leader in energy production and offshore technology, continues to benefit from this groundbreaking innovation.

Though the jet engine and gas turbines were originally developed for aviation, they have since become essential for industrial applications, especially in countries like Norway, where energy production and oil extraction are key parts of the economy. The legacy of Eide's work, along with the continued advancements in turbine technology, has helped shape the modern world in profound ways.

Day 17: England

England has been a major hub of innovation and exploration, especially from the time of the Industrial Revolution. Here are fifteen of the most impactful innovations, explorations, and inventions introduced by the English to the world:

1. **Steam Engine**: Developed by Thomas Newcomen and later improved by James Watt, the steam engine was pivotal in powering factories, trains, and ships during the Industrial Revolution. It transformed industries and transportation worldwide, driving economic and social change.

2. **World Wide Web**: Sir Tim Berners-Lee invented the World Wide Web in 1989 while working at CERN. This invention allowed the internet to become accessible to everyone and revolutionized communication, commerce, and information sharing globally.

3. **Television**: John Logie Baird demonstrated the first working television system in 1925, changing entertainment and information distribution forever. Television became a household staple and a key part of global culture.

4. **Newtonian Physics**: Sir Isaac Newton developed the laws of motion and universal gravitation, forming the foundation of classical mechanics. His work in "Principia Mathematica" had a profound impact on physics, astronomy, engineering, and countless other fields.

5. **Theory of Evolution**: Charles Darwin's theory of evolution by natural selection, outlined in *On the Origin of Species* (1859), revolutionized biology and our understanding of life on Earth. Darwin's work remains foundational in evolutionary biology.

6. **The Industrial Revolution**: The Industrial Revolution began in England in the 18th century, driven by innovations in machinery, textile manufacturing, and transportation. England's developments in industry and labor led to widespread social and economic changes worldwide.

7. **Penicillin**: Discovered by Alexander Fleming in 1928, penicillin was the first widely used antibiotic. This breakthrough led to the

development of antibiotics, saving millions of lives and transforming medicine.

8. **Locomotive Train**: George Stephenson developed the first practical steam-powered locomotive in the 1820s, revolutionizing transportation and enabling faster movement of goods and people, which fueled economic growth.

9. **Magna Carta**: In 1215, the Magna Carta was issued in England as a charter of rights, limiting the powers of the monarchy and laying the foundation for constitutional government. Its principles have influenced democratic systems worldwide.

10. **Radar Technology**: During World War II, British scientists developed radar technology, which was crucial in detecting enemy aircraft. Radar became instrumental in warfare and later in aviation, weather forecasting, and maritime navigation.

11. **Vaccination**: Edward Jenner pioneered the concept of vaccination with his smallpox vaccine in 1796, which led to the eventual eradication of smallpox. Jenner's work laid the foundation for immunology and preventive medicine.

12. **Computer Science (Turing Machine)**: Alan Turing, often considered the father of computer science, conceptualized the Turing machine, a theoretical device that became the foundation of modern computing. His work in artificial intelligence and breaking codes during WWII was revolutionary.

13. **Modern Chemistry**: Sir Humphry Davy and later scientists like Michael Faraday made groundbreaking discoveries in chemistry and electricity, which led to the development of electrochemistry and practical applications in industry.

14. **Genetics and DNA**: Francis Crick, along with James Watson, discovered the double helix structure of DNA in 1953. This discovery transformed biology and medicine, paving the way for advances in genetics, biotechnology, and modern medicine.

These contributions reflect the vast influence of English innovation in shaping the modern world across science, technology, law, medicine, and governance. English discoveries and inventions have continued to advance human understanding, improve quality of life, and drive global progress.

The story of the invention of the Television

The invention of television was a major technological achievement that transformed entertainment and communication worldwide. While many inventors contributed to the creation of television, it was in England that Scottish engineer John Logie Baird made one of the first major breakthroughs. Baird is credited with developing the first working television system and the first publicly demonstrated color television.

Early Vision and Struggles of John Logie Baird

Born in 1888 in Scotland, Baird was fascinated with electronics from a young age. As he pursued a career in engineering, he became increasingly interested in the idea of transmitting images over distance—what we now call television. He faced numerous obstacles, including financial difficulties and health problems, but his persistence led him to experiment with various materials and methods.

Baird's early attempts to create television involved the use of a Nipkow disk, a mechanical device invented by German scientist Paul Nipkow in 1884. The disk was used to scan images and break them into small parts for transmission, which could then be reassembled into a complete image at the receiver's end. However, the technology to make the process practical did not yet exist.

Breakthrough in Mechanical Television

By the early 1920s, Baird moved to London and continued to work on his ideas, experimenting with various mechanical parts and a basic setup. In 1924, he managed to transmit basic shapes and shadows, proving that transmitting images was possible. In 1925, he achieved a significant breakthrough by transmitting the first recognizable human face over his mechanical television system.

In 1926, Baird publicly demonstrated his invention in London at the Royal Institution, showing the first live television images. This was the first time in history that the moving image of a person could be seen and recognized on a screen by an audience. This demonstration is considered the birth of television as a working technology.

Development of Color and Stereo Television

Baird didn't stop at black-and-white television; he was also a pioneer in color television and stereoscopic (3D) television. In 1928, he successfully demonstrated the world's first color transmission using a system with three Nipkow disks and different color filters. He also experimented with 3D television, which laid the groundwork for future innovations in television technology.

Formation of the BBC Television Service

Following Baird's demonstrations, interest in television grew rapidly in the UK. The BBC (British Broadcasting Corporation) began experimental broadcasts in 1930 using Baird's mechanical television system. However, the technology was still limited in terms of image quality and reliability, and it was soon apparent that a fully electronic system would be necessary for television to become mainstream.

Competition with the Electronic System

While Baird pioneered mechanical television, other inventors, notably Philo Farnsworth in the United States and Vladimir Zworykin in the USSR, were developing electronic television systems. By the 1930s, electronic television proved to be superior, as it offered higher resolution, better image stability, and the potential for easier mass production.

Baird attempted to develop his own electronic television system but was ultimately surpassed by companies that adopted fully electronic methods. In 1936, the BBC adopted Marconi-EMI's electronic television system for its broadcasts, which marked the transition from mechanical to electronic television in the UK.

Legacy and Impact of Baird's Invention

Despite mechanical television's limitations, John Logie Baird's contributions laid the foundation for modern television technology. His pioneering work demonstrated the feasibility of transmitting moving images and led to the establishment of the first television broadcasts in the UK.

Baird's influence can still be seen today in the fields of color TV and 3D imaging, both of which he explored long before they became common. Though later surpassed by electronic systems, Baird is remembered as one of television's founding figures.

***

The story of television invention in England is closely tied to John Logie Baird's creativity and determination. His early experiments and public demonstrations showed the world what was possible and inspired further research and development that eventually led to the global phenomenon of television. The BBC's early adoption and the eventual shift to electronic systems ensured that the UK would remain a central player in television broadcasting, contributing to a medium that has shaped society and culture around the world.

The story of the innovation of Vaccination

The invention of vaccination is a landmark in medical history, and its story begins in England in the late 18th century with a country doctor named Edward Jenner. Jenner's work on smallpox inoculation led to the development of the world's first vaccine, transforming public health and laying the foundation for modern immunology.

Background: The Scourge of Smallpox

Before Jenner's time, smallpox was one of the deadliest diseases in the world, with a high mortality rate and serious, often disfiguring, symptoms for survivors. In 18th-century England, smallpox was rampant, affecting nearly everyone at some point in their lives and often leading to death or lifelong scars.

Some societies, including in Asia and Africa, had already practiced variolation—the deliberate introduction of smallpox matter from a mild case of smallpox into a healthy person to induce a mild, protective infection. This practice reduced the severity of smallpox but was risky and still occasionally led to full-blown infections.

Edward Jenner and His Observations

Edward Jenner was born in 1749 in Berkeley, England, and trained as a doctor in London before returning to his hometown to practice medicine. Jenner became aware of a common belief among dairy workers: those who had contracted cowpox—a mild disease affecting cows and humans—seemed to be immune to smallpox.

Intrigued by this observation, Jenner set out to investigate scientifically whether cowpox could indeed protect against smallpox. He hypothesized that exposure to cowpox could act as a safe way to provide immunity against the much more dangerous smallpox.

Jenner's First Experiment with Vaccination

In 1796, Jenner conducted a groundbreaking experiment. He took cowpox material from a lesion on the hand of Sarah Nelmes, a milkmaid infected with cowpox, and inoculated James Phipps, an eight-year-old boy. Phipps developed a mild cowpox infection but recovered quickly.

Weeks later, Jenner exposed Phipps to smallpox material, but the boy did not contract the disease. This demonstrated that cowpox had indeed conferred immunity to smallpox, supporting Jenner's theory. Jenner coined the term "vaccine" from the Latin word *vacca*, meaning "cow," in honor of the role cowpox played in this discovery.

Publication and the Spread of Vaccination

In 1798, Jenner published his findings in a paper titled *An Inquiry into the Causes and Effects of the Variolae Vaccinae*, where he explained his method and described its success. Initially, his work faced skepticism, but he continued to test and document successful vaccinations, eventually convincing the medical community of its effectiveness.

News of Jenner's vaccine spread quickly across England and soon throughout Europe and beyond. Vaccination became widely practiced as an effective and safer alternative to variolation. By 1800, Jenner's vaccine was being administered around the world, and governments, including the British government, endorsed and promoted vaccination as a public health measure.

Opposition and Controversy

Despite its success, vaccination was not without controversy. Some people resisted the idea of introducing material from an animal into the human body, while others had religious objections or feared side effects. Nevertheless, the effectiveness of vaccination in preventing smallpox gradually won over skeptics, and vaccination programs began to be widely implemented.

Global Impact of Jenner's Discovery

Jenner's discovery revolutionized medicine and led to the eventual eradication of smallpox—one of the greatest achievements in public health history. In 1980, after extensive global vaccination efforts, the World Health Organization (WHO) declared smallpox officially eradicated, making it the first human disease to be eliminated through vaccination.

Jenner's work laid the foundation for modern immunology and the development of vaccines for other infectious diseases, ultimately saving millions of lives. Today, vaccination remains one of the most effective ways to prevent diseases, and Jenner is remembered as the "father of immunology."

Edward Jenner's invention of vaccination in England is a story of observation, scientific inquiry, and courage. His discovery changed the course of medical history, leading to a safer world free from the ravages of smallpox. The impact of Jenner's work continues to be felt today, as vaccination remains one of the most powerful tools in global health.

The story of the discovery of the Penicillin

The discovery of penicillin is one of the most important milestones in medical history, marking the beginning of the antibiotic era. This breakthrough occurred in England in 1928 when Scottish scientist Alexander Fleming made a remarkable observation that would ultimately lead to the development of the world's first true antibiotic.

The Accidental Discovery of Penicillin

In 1928, Alexander Fleming was working at St. Mary's Hospital in London. He was studying Staphylococcus bacteria, a common cause of wound infections, hoping to find substances that could inhibit bacterial growth. Fleming was known for his somewhat disorganized work style, and one day he returned from a vacation to find a petri dish in his lab that had accidentally been left uncovered and had grown mold.

Upon closer inspection, he noticed something unusual: wherever the mold had grown on the petri dish, the bacteria around it had been killed. Intrigued, he examined the mold further and identified it as belonging to the Penicillium genus. He soon realized that the mold was producing a substance capable of killing bacteria, which he named penicillin.

Fleming's Initial Research and Challenges

Fleming published his findings in 1929 in the *British Journal of Experimental Pathology*, reporting that penicillin had potent antibacterial properties against a wide range of bacteria. However, he faced significant challenges in isolating penicillin in its pure form. Fleming lacked the resources and expertise to purify and stabilize penicillin for clinical use, so his research on the compound didn't immediately progress further.

Although Fleming recognized the potential of penicillin as a "miracle cure" for bacterial infections, his discovery did not receive widespread attention. He continued his work, but the full potential of penicillin remained unrealized until the early 1940s.

Howard Florey, Ernst Boris Chain, and Mass Production

In the late 1930s, two scientists at the University of Oxford—Howard Florey, an Australian pharmacologist, and Ernst Boris Chain, a German-born biochemist—came across Fleming's 1929 paper and became interested in penicillin's potential. Florey and Chain, along with their research team, decided to pick up where Fleming had left off and work on purifying and producing penicillin.

With funding from the British government and later the United States, the Oxford team successfully purified penicillin and demonstrated its effectiveness in treating bacterial infections in animals. In 1941, they conducted the first clinical trials on human patients with life-threatening infections, and the results were remarkable: penicillin worked as a powerful cure, saving lives that would otherwise have been lost to infections.

The Race to Produce Penicillin During World War II

As World War II intensified, penicillin was in urgent demand for treating wounded soldiers. The British government recognized the importance of penicillin in the war effort and sought help from the United States to mass-produce it. American pharmaceutical companies, with funding and support from the U.S. government, developed large-scale production methods.

By 1944, enough penicillin was available to treat wounded soldiers during the D-Day landings in Normandy, greatly reducing deaths from infected wounds. Penicillin's effectiveness and availability transformed battlefield medicine and helped save countless lives.

Impact of Penicillin

The success of penicillin inspired a surge of research in antibiotics, leading to the discovery of many other life-saving drugs and ushering in the antibiotic era. Fleming, Florey, and Chain were jointly awarded the Nobel Prize in Physiology or Medicine in 1945 for their contributions to the discovery and development of penicillin.

Penicillin marked a turning point in medical history, enabling the treatment of previously deadly bacterial infections and revolutionizing surgery and medicine. It became the model for the development of other antibiotics, fundamentally changing the approach to infectious diseases.

The story of penicillin's invention in England is one of scientific curiosity, accidental discovery, and collaboration. Alexander Fleming's initial discovery of penicillin in 1928 was only the beginning of a journey that would involve many hands to realize its potential. Through the efforts of scientists like Howard Florey and Ernst Chain, penicillin went from a laboratory curiosity to a world-saving drug, and it remains a cornerstone of modern medicine to this day.

The story of the invention of the Steam engine

The invention of the steam engine was a transformative event in history that powered the Industrial Revolution and changed society, transportation, and industry. This story unfolds primarily in England in the 17th and 18th centuries, with contributions from several inventors, each improving upon the work of their predecessors. Key figures in the development of the steam engine include Thomas Savery, Thomas Newcomen, and James Watt, whose innovations ultimately created the modern steam engine.

Early Attempts: Thomas Savery's Steam Pump

The story of the steam engine began in 1698 with Thomas Savery, an English engineer and inventor. Savery was inspired by the problem of flooded mines in England, which needed a more effective method of removing water. Savery designed a device called the "Miner's Friend", which used steam pressure to pump water out of mines.

Savery's machine worked by heating water to produce steam, which created a vacuum as it condensed, drawing water up a pipe. While innovative, Savery's pump had limitations: it was inefficient, could only pump water from shallow depths, and required high pressure, which made it prone to explosions. Nevertheless, Savery's invention was a significant step forward and introduced the concept of using steam to do mechanical work.

Thomas Newcomen's Atmospheric Engine

The next major advancement came from Thomas Newcomen, an ironmonger from Devon, England. Newcomen built on Savery's ideas, developing a more practical steam engine around 1712. His design, known as the Newcomen atmospheric engine, was specifically created to pump water from coal mines.

Newcomen's engine worked by using steam to drive a piston in a cylinder, which was cooled by cold water to create a vacuum and draw the piston down. This up-and-down motion was then used to drive a pump. Newcomen's atmospheric engine was a significant improvement over

Savery's, as it was safer and capable of lifting water from greater depths. His engines were soon widely used in mining operations across England.

Despite its usefulness, Newcomen's engine was inefficient—it required a lot of coal to produce steam and worked slowly. Yet, it was the first practical steam engine, and it laid the groundwork for further innovation.

James Watt and the Birth of the Modern Steam Engine

In the 1760s, a Scottish engineer named James Watt began working on a Newcomen engine at the University of Glasgow and quickly noticed its inefficiencies. Watt realized that much of the energy was wasted in repeatedly cooling and reheating the cylinder, which slowed down the process and consumed excessive fuel.

Watt's breakthrough idea was to add a separate condenser to the engine, which allowed the cylinder to stay hot while the steam was condensed in a separate chamber. This innovation vastly improved efficiency, making the engine faster and more powerful while using significantly less coal.

Watt patented his design in 1769 and later partnered with industrialist Matthew Boulton in 1775 to produce and sell the improved engines. Together, Boulton and Watt transformed the steam engine into a practical power source for a wide range of applications, from mining to manufacturing. Watt's engine was adaptable, could produce rotary motion (essential for machinery), and was highly efficient, making it ideal for powering factories, mills, and, eventually, transportation.

Impact on the Industrial Revolution

Watt's improvements to the steam engine revolutionized industry and laid the foundation for the Industrial Revolution. Factories no longer needed to be near water sources to harness power, and steam engines became the primary power source for manufacturing. As steam engines spread throughout England and beyond, they transformed textile production, ironworks, and other industries, allowing for mass production on an unprecedented scale.

The steam engine also made possible the development of steam-powered transportation, leading to the invention of steamships and steam locomotives. In the early 1800s, engineers like George Stephenson began designing steam engines for railways, and by the 1820s, steam-powered trains were carrying goods and passengers across England. This revolutionized transport, connecting cities, accelerating trade, and shrinking distances, thus contributing to the rapid economic growth of the 19th century.

***

The steam engine's impact extended far beyond industrial production. It changed social structures, sparked urbanization, and led to rapid technological advances in other fields. By the late 19th century, steam power had become an essential part of modern life, driving not only industry but also the expansion of the British Empire through improved transport and communication.

James Watt's innovations in particular laid the foundation for the modern engine and set the stage for subsequent developments in thermodynamics, engineering, and transportation. Today, Watt is often credited as the inventor of the steam engine due to his improvements, and his contributions are commemorated in the unit of power—the watt—named in his honor.

The invention of the steam engine in England was a collaborative process spanning decades, involving several brilliant minds who each advanced the technology toward greater efficiency and practicality. From Thomas Savery's initial concept to James Watt's transformative improvements, the steam engine became the heart of the Industrial Revolution, shaping the modern world and propelling society into a new era of industry, urbanization, and technological progress.

Day 18: France

France has been a center of scientific, cultural, and technological advancement for centuries. Here are fifteen of the most impactful innovations, explorations, and inventions introduced by the French:

1. **Hot Air Balloon**: The Montgolfier brothers, Joseph-Michel and Jacques-Étienne, invented the first successful hot air balloon in 1783. This achievement marked humanity's first successful foray into air travel, sparking interest in aviation and exploration of the skies.

2. **Pasteurization**: Louis Pasteur, a pioneering microbiologist, developed the process of pasteurization to kill harmful bacteria in beverages like milk and wine. This method has had a lasting impact on food safety and storage worldwide.

3. **Photography**: Joseph Nicéphore Niépce and Louis Daguerre were instrumental in developing early photography. Niépce took the first permanent photograph in 1826, and Daguerre introduced the daguerreotype process in 1839, laying the groundwork for modern photography.

4. **The Metric System**: Introduced during the French Revolution, the metric system established a standardized method of measurement that is now the global standard. This system transformed science, trade, and everyday life with its precision and simplicity.

5. **Stethoscope**: French physician René Laennec invented the stethoscope in 1816, revolutionizing medical diagnostics by allowing doctors to listen to patients' heartbeats and lungs without direct contact.

6. **Analytical Geometry**: René Descartes, a French philosopher and mathematician, developed the foundations of analytical geometry, combining algebra and geometry. This system enabled the advancement of calculus and modern mathematics.

7. **Canning for Food Preservation**: Nicolas Appert invented the process of canning food in 1809, which preserved food for extended periods. This innovation was especially important for

long journeys, military expeditions, and later for everyday food storage.

8. **Braille System**: Louis Braille, who was blind himself, invented the Braille system in 1824. This tactile writing system has enabled blind and visually impaired people around the world to read and write independently.

9. **Bicycle**: While not exactly the same as the modern bicycle, French inventors Pierre Michaux and Pierre Lallement designed and built pedal-powered velocipedes in the 1860s, leading to the development of the modern bicycle.

10. **Neon Light**: Georges Claude, a French engineer, invented the neon light in 1910. This technology became iconic for advertising and transformed urban landscapes with bright, colorful signage.

11. **Champagne**: The method of producing sparkling wine, known as the "methode champenoise," was perfected in the Champagne region of France. Dom Pérignon, a Benedictine monk, contributed to this method, making champagne the world-renowned beverage it is today.

12. **Concorde (Supersonic Jet)**: The Concorde, developed by French and British engineers, was the first commercial supersonic passenger jet. It dramatically reduced flight times between continents, marking an era of advanced aviation technology.

13. **Artificial Heart Valve**: French surgeon Alain Carpentier developed the first bioprosthetic heart valve, a significant innovation in cardiac surgery. This device has saved countless lives and advanced cardiovascular treatment.

14. **Theory of Electromagnetism**: André-Marie Ampère, a physicist and mathematician, made key contributions to electromagnetism, laying the groundwork for electrical engineering. His work inspired technologies such as electric motors and generators.

These French contributions reflect France's enduring influence on modern science, technology, culture, and society. From foundational scientific theories and medical advancements to revolutionary inventions in transportation and communication, French innovations continue to shape and inspire the world.

The story of the invention of Pasteurization

The invention of pasteurization was a monumental advance in food safety and preservation, and it came about through the work of the French scientist Louis Pasteur. In the 19th century, Pasteur's discovery of this method, which involves heating liquids to kill harmful microorganisms, revolutionized public health by reducing the spread of diseases through contaminated food and drinks.

The Background: Louis Pasteur's Early Research

Louis Pasteur was born in 1822 in Dole, France, and studied chemistry and microbiology. In the early stages of his career, he focused on studying fermentation, a process used widely in France for making beer, wine, and cheese. Pasteur was called upon by French wine and beer manufacturers to help solve the issue of spoilage, which was causing significant economic losses and was poorly understood at the time.

Through his research, Pasteur discovered that fermentation was caused by microorganisms—yeast, in the case of alcohol production, and different bacteria, which sometimes led to spoilage. He also found that specific bacteria could ruin the wine and beer by producing unpleasant flavors and making the product unsafe for consumption. This was groundbreaking, as it countered the then-prevailing belief in spontaneous generation (the idea that microorganisms arose from non-living matter) and led Pasteur to the idea that microbes in the air were responsible for the spoilage.

Pasteur's Breakthrough with Heating Techniques

By the 1860s, Pasteur began to experiment with heating methods to kill these unwanted bacteria without affecting the taste or quality of the liquid. He discovered that heating the wine or beer to a specific temperature (around 55–60°C or 131–140°F) for a short time could effectively kill most bacteria and fungi, preventing spoilage. Importantly, this gentle heating did not significantly alter the flavor of the product, a crucial factor for wine and beer producers.

Pasteur called this process pasteurization, and he first successfully applied it to wine and beer. The method quickly proved highly effective at preventing spoilage, allowing these beverages to be transported and stored without the risk of bacterial contamination. While pasteurization did not fully sterilize the liquid, it eliminated harmful pathogens and extended the product's shelf life.

Expanding Pasteurization to Milk and Other Foods

Although Pasteur primarily worked with wine and beer, pasteurization was soon applied to milk as well. At the time, milk was a common source of foodborne illnesses, including tuberculosis, brucellosis, and typhoid fever. The discovery that pasteurization could reduce these risks made it a powerful public health tool.

By the 1890s, other scientists and public health officials in Europe and the United States began to advocate for pasteurization of milk. However, it wasn't widely adopted until the early 20th century when public health authorities recognized its potential to prevent widespread diseases. Pasteurization of milk eventually became a standard practice worldwide, helping to reduce the incidence of foodborne illnesses in dairy products.

Importance of Pasteurization

Louis Pasteur's invention of pasteurization had a lasting impact on food safety, medicine, and public health. It also helped solidify his germ theory of disease, which posits that microorganisms are the cause of many illnesses. This theory was revolutionary at the time and led to major advancements in hygiene, sterilization, and disease prevention.

In addition to pasteurization, Pasteur made significant contributions to immunology by developing vaccines for rabies and anthrax. His work laid the groundwork for bacteriology and microbiology, establishing methods that are still used in laboratories today. Pasteur became a national hero in France, and his achievements are celebrated in the field of medicine and science to this day.

The story of pasteurization is one of scientific inquiry, careful experimentation, and practical application. Louis Pasteur's method of heating liquids to kill harmful microbes transformed food preservation and laid the foundation for modern food safety standards. Pasteur's discovery continues to protect people from diseases transmitted through food and drink, marking one of the most impactful contributions to public health in history.

The story of the invention of Photography

The invention of photography was a groundbreaking achievement in visual arts and science, and it began in France in the early 19th century. This process involved several brilliant minds, with key contributions from Nicéphore Niépce and Louis Daguerre. Together, their work led to the development of the first practical photographic processes, capturing permanent images and changing the way humanity records and perceives reality.

The First Photograph: Nicéphore Niépce's Breakthrough

The story of photography began with Joseph Nicéphore Niépce, a French inventor born in 1765. Niépce was fascinated with capturing images from nature, and he experimented with various methods to create permanent images. By 1826, after years of trial and error, he succeeded in producing what is now recognized as the first photograph.

Niépce's process, known as heliography (meaning "sun writing"), involved coating a metal plate with a light-sensitive substance called bitumen of Judea. He placed the plate inside a camera obscura (an early type of pinhole camera) and exposed it to light for several hours. The resulting image, known as *View from the Window at Le Gras*, showed the view from his estate in Saint-Loup-de-Varennes. This photograph was faint but permanent, marking the first successful attempt to capture a real image using light.

Louis Daguerre and the Invention of the Daguerreotype

Niépce knew his heliographic process was limited due to its long exposure times, and he sought a partnership to improve it. In 1829, he teamed up with Louis Daguerre, an artist and theatrical set designer. Daguerre was also fascinated by capturing realistic images, and together, they worked to refine the photographic process.

After Niépce's death in 1833, Daguerre continued the experiments on his own, eventually developing a new process called the daguerreotype. This method used a copper plate coated with silver, which was exposed to iodine vapor to make it light-sensitive. After exposing the plate to light in a camera for a much shorter time (a matter of minutes rather than hours), Daguerre developed the image using mercury vapor, creating a sharp, detailed image.

In 1839, Daguerre publicly announced his invention and presented it to the French Academy of Sciences. The French government recognized the significance of the invention, and in August 1839, the daguerreotype process was gifted "free to the world" by France, allowing anyone to use it without paying for patents.

The Impact and Popularity of the Daguerreotype

The daguerreotype quickly gained popularity, and photography studios began to open across Europe and the United States. People flocked to have their portraits taken, as this new process allowed for realistic, high-quality images that were previously only achievable through painted portraits, which were time-consuming and costly.

The daguerreotype process, however, had some limitations: each image was unique and could not be reproduced, and the metal plate was delicate, requiring careful handling. Nevertheless, it marked the beginning of the photographic industry, and "daguerreotypomania" swept across the world.

Other Innovations and the Development of Modern Photography

At the same time, other inventors were working on new photographic techniques. In 1841, British scientist William Henry Fox Talbot developed the calotype process, which created a paper negative from which multiple positive prints could be made. This advancement in reproducibility was crucial to the evolution of photography, although Talbot's method produced less detailed images than the daguerreotype.

In the decades that followed, photography continued to evolve rapidly. Techniques were developed to create sharper images, shorter exposure times, and eventually color photography. By the late 19th century, photographic technology had advanced significantly, and it had become a popular medium for documenting everything from personal moments to major historical events.

Importance of Photography

The invention of photography transformed visual arts, science, journalism, and personal documentation. Photography allowed for the accurate recording of history, creating a visual archive of the world that extended beyond paintings or written records. It became a powerful tool in journalism, allowing people to see images of events from around the world, and an invaluable resource in fields such as archaeology, medicine, and astronomy.

Photography has also influenced social and cultural life. It allowed ordinary people to preserve personal memories, creating a new kind of family legacy and enabling individuals to see distant parts of the world. Today, photography is embedded in everyday life, from art to communication, and it continues to shape how people interact with the world.

The invention of photography in France began with Nicéphore Niépce's first experiments and culminated in the public unveiling of the daguerreotype by Louis Daguerre. This pioneering work marked the beginning of a new era in visual communication, enabling humanity to capture and share the world in unprecedented ways. The contributions of Niépce and Daguerre, along with other inventors who built on their work, paved the way for modern photography and all the profound changes it has brought to society.

The story of the invention of the Braille system

The invention of the Braille system in France is a remarkable story of resilience, determination, and ingenuity. Created by Louis Braille, a young Frenchman who lost his sight as a child, this tactile writing system allowed blind individuals to read and write independently for the first time. Today, Braille is used worldwide and has empowered countless visually impaired individuals to access education, literature, and communication.

Early Life of Louis Braille and His Loss of Sight

Louis Braille was born on January 4, 1809, in Coupvray, France, a small village near Paris. His father was a leatherworker, and as a young boy, Louis enjoyed playing in his father's workshop. When he was three years old, he accidentally injured his eye with an awl, a sharp tool used for making holes in leather. The wound became infected, and despite medical efforts, the infection spread to his other eye, leaving him completely blind by the age of five.

Education and Encounter with Night Writing

Despite his blindness, Braille was a bright and curious child, and his parents were determined to give him a good education. At age 10, he won a scholarship to the Royal Institute for Blind Youth in Paris, the first school in the world for blind students. The school used a system of raised letters, called embossed letters, which allowed students to feel the shapes of the alphabet. However, it was a slow and inefficient method, as it required students to learn the shapes of letters by touch, which was challenging and time-consuming.

When Braille was 12 years old, he learned about a communication system called night writing developed by Charles Barbier, a captain in the French Army. Barbier had designed this system of raised dots and dashes as a way for soldiers to communicate silently in the dark without using light. Night writing, however, was complex, with symbols representing sounds rather than letters, making it difficult to learn and use.

The Creation of the Braille System

Inspired by Barbier's night writing, Braille set out to create his own simplified system. He wanted a code that would allow blind people to read and write as efficiently as sighted individuals. Over the next few years, he worked tirelessly to develop a system that used a six-dot cell to represent letters and numbers. Each character could be identified by the arrangement of these six dots, allowing for a straightforward and intuitive reading process.

By 1824, at the age of 15, Braille had completed his system, which he called simply Braille. His six-dot code was simple yet versatile, with each dot arrangement representing a letter, number, or punctuation mark. The entire alphabet, numbers, and even musical notation could be represented in Braille's code. The use of six dots also made it possible to fit more symbols onto a page, improving reading speed and efficiency.

Refinement and Promotion of the Braille System

Louis Braille continued to refine and promote his system, which he demonstrated to other students and teachers at the institute. Although the students quickly embraced it, the school administration was initially reluctant to adopt Braille's code, largely due to resistance to change and skepticism about the new system.

Despite this resistance, Braille continued advocating for his code. In 1829, he published the first book explaining his system, *Method of Writing Words, Music, and Plain Songs by Means of Dots, for Use by the Blind and Arranged for Them*. In 1837, he released an updated version that included mathematical symbols and further improvements, solidifying the system's versatility and practicality.

Adoption and Legacy of Braille's System

Louis Braille worked as a teacher at the Royal Institute, where he continued to use and promote his system. However, Braille himself faced many challenges during his lifetime. The administration at his school was slow to adopt the system formally, and he struggled with health issues, particularly tuberculosis, which eventually claimed his life in 1852 when he was just 43 years old.

Although Braille died before seeing his invention gain widespread acceptance, his system gradually gained recognition in France and abroad. By the late 19th century, the Braille system was officially adopted by institutions for the blind throughout Europe and the United States. Today, Braille is a universal reading and writing system for the visually impaired, used in nearly every language and on a wide variety of devices.

Importance and Legacy of the Braille System

The Braille system revolutionized accessibility for the blind and visually impaired, allowing them to participate in education, employment, and society more fully. Braille's code made it possible to read books, take notes, compose music, and engage in independent learning—opportunities that had previously been out of reach.

Modern Braille systems have adapted to include digital Braille displays and advanced assistive technologies, enabling the visually impaired to use computers, smartphones, and other devices. Braille's influence extends beyond reading and writing; his work has inspired ongoing efforts toward accessibility and inclusion for people with disabilities.

The story of Braille's invention is one of resilience and innovation. Louis Braille's six-dot system created a pathway to literacy, independence, and opportunity for millions of visually impaired people around the world. His legacy lives on, and Braille remains an enduring symbol of accessibility, empowerment, and the profound impact one individual can have on society.

The story of the invention of the Stethoscope

The invention of the stethoscope in France is attributed to René Laennec, a French physician whose curiosity and resourcefulness led to one of the most iconic tools in medical history. This simple device, initially created out of necessity, transformed medical diagnostics by allowing doctors to listen to internal sounds of the body, greatly enhancing the ability to diagnose respiratory and cardiac conditions.

The Inspiration: A Doctor's Dilemma in 1816

The story of the stethoscope began in 1816 in Paris, where René Laennec was practicing medicine at the Necker-Enfants Malades Hospital. Laennec, known for his deep interest in understanding chest diseases, often relied on a common diagnostic method called percussion (tapping the chest and listening to sounds) and auscultation (listening directly by placing the ear on the patient's chest). However, this method posed challenges, especially with certain patients, such as women, where physical contact might be considered improper, or with patients where listening closely was ineffective.

One day, Laennec was faced with a young woman presenting with possible heart disease. Feeling uncomfortable placing his ear directly on her chest, he recalled an experiment he'd seen in which sound could be amplified through wood. This gave him an idea. He rolled up a piece of paper into a tube and placed it against the woman's chest and his ear. To his surprise, the sounds were clearer than he'd ever heard with direct auscultation.

The Development of the Stethoscope

Laennec immediately recognized the potential of this approach. Soon after, he began experimenting with a variety of materials to create a more durable version of the rolled paper tube. He ultimately crafted a hollow, wooden cylinder, approximately one inch in diameter and a foot long, which became the first true stethoscope.

This device allowed Laennec to hear the sounds of the lungs and heart in much greater detail. For the first time, he could distinguish different respiratory and cardiac sounds, which enabled him to identify and describe specific conditions such as bronchitis, pneumonia, and pleurisy with remarkable accuracy. He referred to this practice as mediate auscultation, as opposed to direct auscultation, which involved placing the ear directly on the body.

Publication and the First Medical Text on Auscultation

In 1819, after three years of careful observation and refinement of his techniques, Laennec published his findings in a book titled *De l'Auscultation Médiate* (On Mediate Auscultation). This book was the first comprehensive text on auscultation, detailing the sounds associated with various chest diseases and introducing the use of the stethoscope in clinical practice. Laennec's work laid the foundation for modern pulmonology and cardiology, as he provided the first thorough descriptions of how different sounds could indicate specific diseases.

Initial Resistance and Later Adoption of the Stethoscope

Although the stethoscope proved effective, it was not widely accepted at first. Some doctors found the device cumbersome, while others were skeptical about its practical benefits. However, as Laennec's students and colleagues began to adopt it, the stethoscope's value in diagnosing chest conditions became undeniable.

Over the next few decades, the stethoscope underwent modifications to make it more user-friendly. In the 1850s, a French physician named Pierre Piorry added an earpiece, and later, in the United States, George Cammann designed a version with two earpieces (binaural stethoscope), making it more convenient and effective. This binaural model became the standard and remains the basis for modern stethoscope design.

Importance of the Stethoscope

The stethoscope fundamentally changed medicine by enhancing diagnostic accuracy and deepening physicians' understanding of the human body. With Laennec's invention, doctors could now listen to subtle changes in internal sounds, detect early stages of disease, and monitor the progression of conditions in a non-invasive way. The stethoscope quickly became an essential diagnostic tool, symbolizing the medical profession and even giving rise to the phrase "wearing the stethoscope" to mean practicing medicine.

Today, the stethoscope remains a symbol of healthcare and is still widely used, despite advances in diagnostic imaging. Digital stethoscopes, with features such as sound amplification and data recording, build on Laennec's legacy, blending tradition with technology.

René Laennec's invention of the stethoscope is a story of resourcefulness and innovation. His simple solution to a practical problem led to an instrument that has saved countless lives by enabling early diagnosis and improved patient care. The stethoscope remains a testament to how a small invention, inspired by a modest need, can revolutionize an entire field and become an enduring symbol of compassionate and attentive medical care.

Day 19: Spain

Spain has contributed significantly to global knowledge and technology through its innovations, explorations, and inventions in various fields, including navigation, medicine, and cultural arts. Here are fifteen of the most impactful contributions introduced by the Spanish to the world:

1. **Global Maritime Exploration**: Spain spearheaded the Age of Exploration, financing expeditions that led to the discovery of new continents. Explorers like Christopher Columbus, Ferdinand Magellan, and Juan Sebastián Elcano (the first to circumnavigate the globe) expanded global understanding of geography, establishing trade routes and cultural exchanges.

2. **Astronomical and Mathematical Innovations**: Spanish-Arabic scholars such as al-Zarqali (Arzachel) contributed to early advancements in astronomy and mathematics. His work, *The Tables of Toledo*, influenced astronomical knowledge in Europe, introducing more accurate calculations of planetary movements.

3. **Sherry and Wine Innovations**: Spain's regions, particularly Jerez, are known for pioneering wine production techniques that led to the creation of sherry, which became a highly prized export. Spain's influence in winemaking also helped shape vineyard practices in Latin America and beyond.

4. **Antibiotics (Isolation of Streptomycin)**: Although streptomycin was formally discovered later in the U.S., Spanish bacteriologist Juan de la Cierva y Hoces contributed significantly to the antibiotic field by laying the foundation for isolating antibiotic-producing bacteria, impacting medicine worldwide.

5. **The Guitar**: Although guitars have ancient origins, Spanish luthiers in the 19th century, such as Antonio de Torres Jurado, refined the modern classical guitar design. Spain's influence helped make the guitar one of the world's most popular musical instruments.

6. **Flamenco Music and Dance**: This art form, originating in Andalusia, blends influences from Spanish, Romani, and other cultures. Flamenco music, dance, and rhythms have become iconic

symbols of Spain and have greatly influenced music and dance globally.

7. **The Space Suit**: Spanish engineer Emilio Herrera developed the "escafandra estratonáutica," a type of pressurized suit in the 1930s that was intended for high-altitude flights. His design later influenced the development of modern space suits used by astronauts.

8. **Gyroplane**: Spanish engineer Juan de la Cierva invented the autogyro (gyroplane) in 1923, a precursor to the helicopter. His invention allowed for safer flight at slower speeds and laid the foundation for the development of rotary-wing aircraft.

9. **Chocolate as a Confection**: After the Spanish encountered cacao in the Americas, they adapted it, adding sugar to make it a sweet treat. Spain introduced chocolate to Europe, leading to the confectionary revolution and the popularity of chocolate worldwide.

10. **Microsurgery Techniques**: Spanish ophthalmologist Ignacio Barraquer developed pioneering techniques in microsurgery, particularly in cataract surgery, that were highly influential and continue to benefit patients worldwide.

11. **Surgical Advances (The Autotransfusion Device)**: Spanish physician Dr. Jordi Caralps pioneered methods in trauma surgery, including the autotransfusion device, which became essential during wartime and was widely adopted in emergency medicine.

12. **Chupa Chups Lollipops**: Created in the 1950s by Spanish entrepreneur Enric Bernat, Chupa Chups popularized the lollipop format and became one of the most recognizable candy brands worldwide, symbolizing Spanish creativity in confectionery.

These contributions illustrate Spain's rich heritage in exploration, science, and the arts. Spanish innovations and cultural exports continue to have an enduring impact on global society and history, from advancements in aviation and medicine to the introduction of culinary and artistic traditions.

The story of the invention of the Autotransfusion device

The autotransfusion device was a pioneering development in Spain that transformed emergency and surgical care by allowing patients to safely receive their own blood during surgery or after trauma. The invention is credited to Dr. Jordi Caralps, a Spanish surgeon, in the 1970s. Driven by the limitations and risks of traditional blood transfusions, Dr. Caralps set out to develop a method that would minimize blood loss and make surgeries safer.

The Need for an Autotransfusion Device

Historically, during surgery or traumatic injuries, significant blood loss often required blood transfusions from external donors. While blood transfusions were lifesaving, they also carried risks, including potential infections, immune reactions, and compatibility issues. This problem was particularly critical in Spain, where, at the time, blood donations were sometimes insufficient to meet the demand.

Dr. Jordi Caralps, a prominent surgeon, saw this need firsthand in the operating room. He envisioned a device that could collect a patient's own blood lost during surgery, clean it, and reinfuse it back into the same patient. This process, called autotransfusion, offered several advantages: it avoided the need for donor blood, reduced the risk of immune reactions, and ensured compatibility.

Development of the Autotransfusion Device

In the early 1970s, Dr. Caralps began researching and developing a device that could safely recycle a patient's blood during surgery. His goal was to create a machine that could suction blood lost during an operation, filter out debris, separate plasma from red blood cells, and return the purified blood to the patient.

After rigorous testing, he successfully developed a working prototype of the autotransfusion device. The device collected and cleaned the blood, making it suitable for reinfusion. This innovation greatly reduced the need for external blood transfusions, and it became particularly useful in major surgeries, such as cardiac or orthopedic operations, where blood loss could be substantial.

Clinical Trials and Success

After developing the prototype, Dr. Caralps conducted clinical trials to test its safety and effectiveness. These trials showed promising results, and the autotransfusion device quickly proved to be a revolutionary tool in surgeries with high blood loss. Spanish hospitals began to adopt the device, and it gained international attention for its potential to improve patient outcomes and reduce reliance on blood donations.

Impact and Global Adoption

Dr. Caralps' invention soon spread beyond Spain, influencing surgical practices worldwide. The device was particularly valuable in settings where donor blood was scarce or risky to use, as it allowed for more controlled and safer blood management. Its introduction contributed significantly to the development of modern blood management strategies and inspired further advancements in autotransfusion technology.

The Legacy of the Autotransfusion Device

Today, autotransfusion devices are an integral part of surgical and trauma care. Dr. Caralps' work laid the groundwork for modern devices, which are now more compact and capable of processing blood in real-time. His invention has saved countless lives by reducing the risks associated with traditional blood transfusions and optimizing the use of a patient's own blood.

The story of Dr. Jordi Caralps and his autotransfusion device is a testament to innovation driven by the desire to improve patient care. His invention transformed surgical medicine, and its impact continues to be felt in hospitals around the world.

Day 20: Morocco

Morocco, with its unique geographic position at the crossroads of Africa, Europe, and the Middle East, has a long history of innovations and contributions that have impacted the world in fields such as science, exploration, and culture. Here are ten important innovations, explorations, and inventions introduced by Moroccans:

1. **World Atlas by Al-Idrisi**: In the 12th century, Moroccan geographer and cartographer Muhammad al-Idrisi created one of the most detailed world maps of his time, known as the *Tabula Rogeriana*. This map was groundbreaking for its accuracy and greatly advanced European and Middle Eastern knowledge of geography.

2. **Agricultural Innovations and Irrigation**: The Moors, many of whom hailed from the Maghreb, including Morocco, brought advanced irrigation systems and agricultural techniques to Spain. This knowledge helped transform agriculture in the Iberian Peninsula, allowing for more diverse crops and efficient water use.

3. **Marinid Medersa Architecture**: Morocco developed unique architectural styles, especially during the Marinid dynasty (13th-15th centuries). Medersas (Islamic schools) such as the Bou Inania and Al-Attarine in Fes and Meknes exemplified intricate geometric designs and influenced Islamic architecture throughout North Africa and the Islamic world.

4. **Introduction of Citrus Fruits to Europe**: The introduction of citrus fruits like oranges and lemons to Europe is largely credited to the trade routes established by the Moors in Al-Andalus (modern-day Spain and Portugal). These fruits became central to European agriculture and cuisine.

5. **Advanced Astronomy**: In the 14th century, Moroccan astronomers and scholars such as Ibn Battuta contributed to the development of astronomy in North Africa, often working with

astronomical devices and instruments like the astrolabe, which was essential for navigation and timekeeping.

6. **Islamic Calligraphy and Manuscripts**: Morocco became a center for the art of Islamic calligraphy, particularly the Maghrebi script. The country developed unique calligraphic styles, contributing significantly to the visual arts and the preservation of Islamic manuscripts.

7. **Mint Tea (Atay)**: Moroccan mint tea, also called "atay," has become one of Morocco's most enduring cultural exports. It originated from a fusion of Arab and Berber traditions and has since spread across North Africa and the Middle East, becoming a symbol of Moroccan hospitality.

8. **The Riffian Berber Musical Traditions**: Moroccan music, particularly that of the Riffian Berbers, developed unique styles and instruments, such as the ribab (a bowed lute) and the guembri (a three-stringed instrument). These musical styles have influenced North African and world music, contributing to genres such as Gnawa and Andalusian music.

9. **Maritime Exploration by Ahmad al-Mansur**: Sultan Ahmad al-Mansur of Morocco promoted exploration and trade with Sub-Saharan Africa in the late 16th century, particularly during the Saadian Dynasty. His expeditions to the Niger River Valley expanded Morocco's trade networks, extending influence and commerce deeper into the African continent.

These Moroccan contributions demonstrate the country's impact on fields like geography, architecture, agriculture, and the arts. Morocco's history of innovation, exploration, and cultural influence has left a lasting legacy on North Africa, Europe, and beyond.

The story of the innovative creation of the World Atlas

The creation of the *Tabula Rogeriana*, or *Book of Roger*, by the Andalusian-born geographer Al-Idrisi is one of the most remarkable stories of medieval cartography. This world atlas was commissioned by King Roger II of Sicily and completed in 1154. Although the project took place in Sicily, Al-Idrisi's work has connections with Morocco and the broader Islamic world, as he was deeply influenced by his Andalusian-Moroccan roots and the rich tradition of Islamic scholarship. Here's the story of how this groundbreaking atlas came to be:

1. Background of Al-Idrisi

Al-Idrisi, whose full name was Abu Abdullah Muhammad al-Idrisi, was born in Ceuta, a city that sits at the northern tip of Morocco, likely around 1100. Ceuta was then part of the Almoravid Empire, which connected Al-Idrisi with both the Islamic scholarship of Al-Andalus (modern Spain and Portugal) and the rich intellectual heritage of the Islamic world. As a young man, he traveled extensively across North Africa, the Middle East, and Europe, learning from scholars in different regions and gathering geographic knowledge.

His travels provided him with invaluable information about different lands and peoples, setting the foundation for his later work in cartography and geography.

2. Invitation to Sicily

King Roger II of Sicily, a Norman ruler with a keen interest in science and knowledge, heard of Al-Idrisi's skills and invited him to his court in Palermo, Sicily. Roger was an unusually open-minded Christian king, eager to foster a multicultural, multilingual society that brought together Arab, Greek, and Latin traditions. He sought to compile the most accurate depiction of the known world, surpassing the knowledge held by Christian and Muslim scholars of the time.

Al-Idrisi accepted the invitation and moved to Palermo, where he joined Roger's court and set out on a project that would occupy him for years: the creation of a world atlas.

3. Research and Data Collection

Al-Idrisi did not rely solely on his travels to create his maps; he conducted extensive research using sources from various Islamic, Greek, Roman, and even Persian texts. Additionally, Roger's court gathered information from travelers, merchants, and explorers who visited Sicily, contributing insights from many distant regions.

Al-Idrisi cross-referenced these sources meticulously, comparing different accounts and correcting inaccuracies. This level of detail was uncommon in medieval European maps, which often included more myth than accuracy. Al-Idrisi, however, was deeply committed to a scientific approach to geography.

4. Compiling the Atlas

The project took about 15 years to complete. Al-Idrisi worked with a team of artisans and craftsmen to produce the *Tabula Rogeriana*, a massive silver disc engraved with a map of the world, along with an accompanying text describing different regions, peoples, and geographic features. The map was oriented with the south at the top, which was common in Islamic cartography.

The atlas included detailed descriptions of regions as far as Scandinavia, Western Europe, North Africa, the Middle East, and parts of Asia. Al-Idrisi's atlas was far more accurate and comprehensive than any previous European or Islamic maps, incorporating both topographical details and descriptions of local cultures and economies.

5. The *Tabula Rogeriana* (Book of Roger)

The resulting work, *Kitab Rujar* (Book of Roger), was completed in 1154. It contained a highly detailed world map divided into 70 sections, with each section accompanied by a description. This arrangement allowed readers to focus on individual regions while still understanding their place within the broader world.

Al-Idrisi's work included precise measurements and descriptions of cities, rivers, mountains, and even trade routes. His work also dispelled some of the misconceptions of his time, such as the idea that the Indian Ocean was a landlocked sea.

6. Al-Idrisi's Influence on Geography

Al-Idrisi's *Tabula Rogeriana* became one of the most influential works of medieval geography, studied by scholars in both the Islamic world and Europe. Though the original silver map is lost, several manuscript copies of his work survived, preserving his geographic knowledge. His work remained a standard reference in both the Islamic and Christian worlds for several centuries and influenced later cartographers during the Renaissance.

***

Though Al-Idrisi completed his atlas in Sicily, his legacy is deeply connected to Morocco and the Islamic world. His work exemplifies the richness of Islamic scholarship during the medieval period, which valued empirical knowledge, scientific rigor, and a multicultural approach. Al-Idrisi's upbringing and education in Ceuta, a city rich in Moroccan-Andalusian culture, were instrumental in shaping his worldview and fostering his curiosity.

In summary, Al-Idrisi's *Tabula Rogeriana* was more than just a map; it was a groundbreaking synthesis of geographic knowledge from across cultures. It demonstrated how a single project could unite knowledge from different parts of the world and stand as a testament to human curiosity and cross-cultural collaboration.

Day 21: USA

The United States has been at the forefront of innovation and exploration, particularly since the Industrial Revolution. Here are 25 of the most impactful American contributions in science, technology, medicine, and other fields:

1. **The Light Bulb**: Although electric lighting concepts existed, Thomas Edison's practical and long-lasting light bulb in 1879 made widespread indoor lighting a reality and changed modern life.

2. **The Airplane**: In 1903, the Wright brothers achieved the first controlled, sustained flight with a powered aircraft, transforming transportation and paving the way for the aviation industry.

3. **The Internet**: The internet was originally developed by DARPA (Defense Advanced Research Projects Agency) in the 1960s and 1970s. It has revolutionized communication, information sharing, and commerce globally.

4. **Personal Computer**: In the 1970s and 1980s, American companies like Apple and IBM pioneered the personal computer, making computing accessible to the general public and revolutionizing workplaces, education, and households.

5. **The Assembly Line**: Henry Ford revolutionized manufacturing in the early 20th century by implementing the assembly line, significantly reducing the cost and time of production and making goods more affordable.

6. **Nuclear Power**: The USA developed nuclear energy, first for military purposes with the atomic bomb and later for civilian energy production, providing a new and powerful energy source.

7. **Polio Vaccine**: Dr. Jonas Salk developed the first effective polio vaccine in the 1950s, virtually eradicating polio in many parts of the world and saving millions of lives.

8. **The Global Positioning System (GPS)**: Initially developed by the U.S. Department of Defense, GPS technology enables precise location tracking and has countless applications in navigation, logistics, and personal devices.

9. **Space Exploration (Apollo Moon Landing)**: NASA's Apollo 11 mission in 1969 made the USA the first nation to land humans on the Moon, marking a monumental achievement in space exploration.

10. **Silicon Microchip**: The integrated circuit, or microchip, developed in the U.S. in the late 1950s by Jack Kilby and Robert Noyce, is the basis for modern electronics, enabling the development of computers, phones, and many other digital devices.

11. **The Laser**: Developed in the early 1960s, the laser has applications in medicine, telecommunications, manufacturing, and consumer electronics, transforming industries and daily life.

12. **Digital Camera**: Invented at Eastman Kodak in 1975 by Steven Sasson, the digital camera set the stage for modern digital photography and media, significantly impacting social media, journalism, and personal communication.

13. **Transistor**: Developed in 1947 by American physicists John Bardeen, William Shockley, and Walter Brattain, the transistor is foundational to modern electronics, leading to the creation of compact, affordable computers and devices.

14. **Social Media Platforms**: American companies like Facebook, Twitter, and YouTube introduced social media, transforming communication, social interactions, news dissemination, and global connectivity.

15. **3D Printing**: This additive manufacturing technology, invented in the 1980s by Charles Hull, allows objects to be created layer by layer and has applications in fields as diverse as medicine, engineering, and space exploration.

16. **Modern Air Conditioning**: Willis Carrier invented the first modern air conditioner in 1902, transforming living and working environments and enabling the development of industries in warmer climates.

17. **ATM (Automated Teller Machine)**: The first successful ATM was developed in the U.S. in 1969, allowing people access to cash at any time and revolutionizing banking convenience.

18. **Email**: Ray Tomlinson, an American engineer, created the first email system in 1971, pioneering a method of digital communication that has become essential for business and personal communication.

19. **Credit Card**: The first modern credit card was introduced by American companies in the 1950s, transforming consumer spending and leading to the development of the global credit and payment processing industries.

20. **Video Games**: Pong, created by Atari in 1972, was one of the first arcade video games and popularized video gaming, leading to the massive entertainment industry we have today.

21. **MRI (Magnetic Resonance Imaging)**: Although based on earlier international research, the first practical MRI machine was developed in the U.S. in the 1970s. MRI revolutionized medical imaging and diagnostics, offering detailed internal images without surgery.

22. **Biotechnology (Recombinant DNA)**: The development of recombinant DNA technology by American scientists in the 1970s launched the biotech industry, allowing for genetic engineering, pharmaceuticals, and agricultural innovations.

23. **Electric Car (Modern Era)**: Tesla Motors, an American company, popularized the electric car in the 21st century by making electric vehicles efficient, stylish, and desirable, pushing the auto industry toward sustainable alternatives.

24. **CRISPR Gene Editing**: CRISPR technology, developed with major contributions from American scientists, enables precise editing of DNA and holds promise for treating genetic diseases, revolutionizing genetic science and medicine.

These innovations underscore the USA's pivotal role in transforming technology, healthcare, communication, and the environment. Many of these inventions not only changed daily life but also set the stage for future advancements worldwide.

The story of the invention of the Internet

The invention of the internet began as a U.S. government initiative to improve communication technology, security, and data-sharing capabilities, especially for military purposes. Here's how this groundbreaking innovation evolved from an experimental project to the global network we use today:

The Cold War Context and ARPA's Creation

In the late 1950s, during the Cold War, the U.S. was concerned about maintaining a technological edge over the Soviet Union, especially after the USSR launched Sputnik in 1957.

In response, the U.S. government established the Advanced Research Projects Agency (ARPA), later known as the Defense Advanced Research Projects Agency (DARPA), in 1958. ARPA was tasked with fostering advanced technological research that could bolster national security.

The Concept of a Decentralized Network

ARPA wanted a communication system that could withstand attacks and remain operational even if parts of the network were destroyed. This led to the concept of a decentralized network, which would be resilient in the event of physical attacks or disasters.

Paul Baran at RAND Corporation and Donald Davies at the National Physical Laboratory in the UK both independently proposed the idea of packet switching in the early 1960s. Packet switching breaks data into smaller chunks, or packets, and sends them independently across a network, allowing for more efficient and reliable transmission.

Birth of ARPANET (1969)

In 1966, ARPA (now DARPA) launched the ARPANET project, which aimed to connect computers from different universities and research institutions.

The team at ARPA, led by Lawrence Roberts, started collaborating with computer scientists to implement a packet-switched network. This led to the first successful connection between two computers: a link between UCLA and Stanford Research Institute (SRI) in October 1969.

The first message sent was supposed to be "LOGIN," but only "LO" was transmitted before the system crashed. Still, it marked the first packet-switched communication, and ARPANET gradually expanded to include more nodes.

Development of Network Protocols: NCP and TCP/IP

In the early 1970s, the Network Control Protocol (NCP) was developed as ARPANET's initial communications protocol.

However, as ARPANET grew, limitations of NCP became clear, especially as new networks emerged. Researchers needed a standardized protocol that would allow different types of networks to communicate.

In 1973, Vint Cerf and Bob Kahn proposed the Transmission Control Protocol (TCP), later combined with the Internet Protocol (IP), creating TCP/IP. TCP/IP became the foundational protocol for the internet, enabling communication between different networks and setting the stage for the global connectivity we see today.

Expansion Beyond the Military: NSFNET

By the 1980s, the concept of networked communication had gained traction beyond the military. The National Science Foundation (NSF) established NSFNET in 1986 to connect supercomputing centers across the U.S. and provide internet access to universities and research institutions.

NSFNET helped lay the groundwork for a public, global internet by connecting academic institutions and allowing for data exchange on a broader scale.

Commercialization and the World Wide Web

In 1989, British scientist Tim Berners-Lee invented the World Wide Web at CERN, developing a system of hypertext to enable easier information sharing. This allowed users to navigate between linked documents, setting the foundation for modern websites.

In 1991, the World Wide Web was released to the public, and in 1993, the first popular web browser, Mosaic, made it easier to access and view web pages. The internet's potential for business, entertainment, and education became apparent, and commercial internet service providers (ISPs) began emerging.

The Internet Boom of the 1990s

The U.S. government privatized the internet in the early 1990s, allowing commercial entities to build internet infrastructure. This led to rapid growth, with businesses and households beginning to connect to the internet.

The rise of email, e-commerce, search engines, and social networking in the late 1990s further fueled the internet's explosive growth and its integration into everyday life.

***

From its origins as a military research project, the internet has transformed into a global platform for communication, commerce, and knowledge-sharing.

The fundamental ideas behind its decentralized design, open protocols like TCP/IP, and a commitment to global standards have allowed the internet to expand and evolve, connecting billions of people worldwide and enabling revolutionary changes across all areas of life.

The internet, born out of the vision of American researchers and agencies, represents a quintessential example of collaboration, curiosity, and innovation. Its impact is both vast and continuously evolving, with new applications and technologies building on this foundational achievement every day.

The story of the invention of the Airplane

The invention of the airplane is credited to Orville and Wilbur Wright, two brothers from Dayton, Ohio, who made history with the first powered, controlled, and sustained flight in 1903. Here's the story of their groundbreaking invention and how it transformed transportation and technology forever:

The Wright Brothers' Early Curiosity and Bicycle Business

Born in the 1860s, Orville and Wilbur Wright were inspired by their love for mechanical innovation. As young men, they operated a successful bicycle repair and manufacturing business in Dayton, Ohio, which allowed them to hone their mechanical skills and fund their aviation experiments.

They were particularly fascinated by the concept of flight, having read about German aviation pioneer Otto Lilienthal and observing how birds used their wings to control motion.

Inspired by Gliders and Early Research

The Wright brothers began by studying the principles of aerodynamics and reviewing the work of pioneers like Lilienthal, Samuel Langley, and Octave Chanute. They believed that controlled, powered flight was achievable and that human skill in manipulating an aircraft would be essential.

In 1899, they constructed a small biplane kite to test their ideas on wing-warping, a technique they devised to control the roll of the aircraft by twisting or "warping" the wings. This idea would prove crucial to controlling an aircraft.

Practical Experiments in Kitty Hawk

The Wrights chose Kitty Hawk, North Carolina, as their testing grounds due to its steady winds and sand dunes, which provided a soft landing. They began experimenting with gliders in 1900, improving their designs and understanding how to control pitch, yaw, and roll.

Over the next few years, they meticulously studied and refined their gliders, developing methods to control an aircraft's direction and stability by adjusting wing shape, and they created a movable tail to help control yaw (left-to-right movement).

Building a Powered Flying Machine

By 1903, the Wrights were ready to move beyond gliders. They designed and built their own lightweight engine with the help of their bicycle shop mechanic, Charles Taylor, since no engine manufacturers could meet their specifications.

The engine was relatively small but powerful enough for their purpose, and they connected it to a propeller system of their own design. The Wrights had become the first to accurately calculate the required propeller shapes and speeds for flight.

The First Powered Flight on December 17, 1903

On December 17, 1903, the Wright brothers were ready to test their new aircraft, the Wright Flyer. The plane had a wingspan of 40 feet, was made of wood and fabric, and had a 12-horsepower engine driving two propellers.

Orville took the first flight, staying airborne for 12 seconds and covering a distance of 120 feet. Later that day, Wilbur flew the longest flight, covering 852 feet in 59 seconds. This was the first controlled, sustained, and powered flight in history.

Further Development and Public Demonstrations

After their success, the Wrights continued to improve their aircraft and flying techniques. They returned to Dayton and conducted further tests, improving stability and control.

In 1908, they took their flights public, demonstrating their airplane's capabilities in the United States and France. These demonstrations attracted worldwide attention and convinced skeptics that powered flight was achievable and commercially viable.

Patents, Legal Battles, and Business Ventures

The Wright brothers obtained a patent for their flight control system in 1906. However, they were involved in legal battles over their patents as other inventors attempted to develop their own aircraft, often using similar control methods.

Eventually, they started the **Wright Company** in 1909, manufacturing airplanes and training pilots. Their designs set the standard for aviation and established the brothers as leaders in the field.

***

The Wright brothers' achievements inspired further developments in aviation. Their methods of wing-warping and control were replaced by more advanced techniques, but their fundamental discoveries laid the groundwork for modern aerodynamics and aviation engineering.

The Wright Flyer became an iconic symbol of innovation, and the brothers are celebrated as pioneers who made air travel and modern flight possible.

The invention of the airplane revolutionized travel, commerce, and warfare. It opened up new possibilities for global connectivity, led to the creation of the aerospace industry, and fundamentally changed transportation. The Wright brothers' meticulous approach to experimentation, control, and design became a model for scientific and engineering innovation, marking one of the most important technological advancements in history. Their legacy lives on as a testament to perseverance, creativity, and the transformative power of flight.

The story of the invention of the Polio Vaccine

The story of the polio vaccine's invention is one of the most significant breakthroughs in medical history. It brought an end to one of the most feared diseases of the 20th century and saved countless lives. This success was largely driven by Dr. Jonas Salk and later by Dr. Albert Sabin, who developed two different types of polio vaccines that led to the near eradication of the disease.

The Polio Epidemic and Fear in America

Polio, or poliomyelitis, is a highly infectious viral disease that can cause paralysis or even death. Outbreaks swept through communities, especially in the early to mid-1900s. It primarily affected children and often left survivors with lifelong disabilities.

The summer months saw the highest infection rates, with public places like pools, parks, and theaters avoided for fear of contracting the virus. This created widespread panic, as the disease seemed unstoppable and unpredictable.

Early Research and the National Foundation for Infantile Paralysis

Efforts to combat polio were organized by Franklin D. Roosevelt, who was himself paralyzed by the disease. In 1938, Roosevelt helped establish the National Foundation for Infantile Paralysis, later known as the March of Dimes, which funded research and provided care for polio victims.

The organization mobilized public support, raising millions of dollars for research and treatment. This funding created a strong foundation for scientists to study the virus and eventually develop a vaccine.

Dr. Jonas Salk and the Inactivated Polio Vaccine (IPV)

Dr. Jonas Salk, an American virologist, was working at the University of Pittsburgh in the early 1950s and was deeply focused on developing a vaccine. Instead of using a live virus, he chose to use a "killed" (inactivated) form of the virus, known as the **inactivated polio vaccine (IPV)**.

Salk believed that by inactivating the virus with formaldehyde, he could create a vaccine that would stimulate the immune system to produce antibodies without causing illness. His approach was cautious but promising, as it reduced the risk of the vaccine itself causing polio.

The Historic Vaccine Trials of 1954

By 1952, Salk had preliminary results showing that his vaccine was effective in small tests, so he conducted a large-scale trial to prove its efficacy. With funding and support from the March of Dimes, he embarked on the largest medical trial in history.

In 1954, 1.8 million children in the United States participated in the **double-blind field trials** for the vaccine. This monumental effort involved administering the vaccine to some children and a placebo to others, while neither participants nor administrators knew who received which.

The results, announced in April 1955, showed that the Salk vaccine was safe and 80-90% effective at preventing polio.

Widespread Distribution and Impact of the Salk Vaccine

Following the successful trials, the Salk vaccine was quickly approved, and vaccination campaigns were launched across the United States and around the world.

Cases of polio plummeted within a few years of the vaccine's release. Salk became a national hero, and his choice not to patent the vaccine allowed it to be widely distributed, bringing polio outbreaks under control.

Dr. Albert Sabin and the Oral Polio Vaccine (OPV)

While Salk's vaccine was highly effective, it required injections, which posed logistical challenges, especially in developing countries. In the late 1950s, Dr. **Albert Sabin** developed an **oral polio vaccine (OPV)** using a weakened but live virus.

The OPV was taken by mouth, which made it easier to administer and more affordable, particularly in mass vaccination campaigns. Sabin's vaccine also had the added benefit of providing immunity in the gut, which helped prevent the spread of the virus more effectively.

After extensive testing, the OPV was approved for use in the early 1960s. It was used in mass vaccination campaigns that successfully eliminated polio in most parts of the world.

Global Polio Eradication Initiatives

The success of the Salk and Sabin vaccines spurred international efforts to eradicate polio. In 1988, the **Global Polio Eradication Initiative** was launched by organizations like the World Health Organization, UNICEF, and Rotary International.

These campaigns, which relied heavily on the OPV, drastically reduced polio cases worldwide. Today, only a few countries continue to report cases, and efforts continue to eradicate the virus completely.

Legacy of the Polio Vaccine

The invention of the polio vaccine is celebrated as a turning point in public health. It demonstrated the power of vaccination to prevent disease on a massive scale and showed how coordinated efforts could control, and even potentially eliminate, infectious diseases.

Jonas Salk's and Albert Sabin's contributions remain central to vaccine science, and their work helped shape the development and distribution of vaccines worldwide.

The development of the polio vaccine was a milestone in medicine, bringing hope to families and communities affected by polio. It paved the way for future vaccination programs and underscored the importance of collaborative, publicly funded health initiatives in combating infectious diseases.

The story of the invention of the MRI

The invention of Magnetic Resonance Imaging (MRI) revolutionized medical diagnostics, enabling detailed imaging of the body's internal structures without surgery or radiation. The story of MRI's development spans several decades, involving contributions from multiple scientists in the fields of physics, engineering, and medicine. Here's an overview of how MRI came to be:

Foundations in Nuclear Magnetic Resonance (NMR)

MRI's story begins with **nuclear magnetic resonance (NMR)**, a concept first discovered in 1946 by American physicists **Felix Bloch** at Stanford University and **Edward Purcell** at Harvard University.

They discovered that atomic nuclei emit signals when exposed to magnetic fields and radio waves. This phenomenon, known as NMR, allowed scientists to analyze chemical structures by observing how different atoms interacted with magnetic fields.

Bloch and Purcell received the Nobel Prize in Physics in 1952 for their work, which laid the scientific foundation for MRI.

NMR Applied to Medicine

Throughout the 1950s and 1960s, NMR was primarily used in chemistry and physics research to study molecular structures. However, some researchers believed it could be adapted to image biological tissues.

In the 1970s, Dr. Raymond Damadian, a medical doctor and researcher in the United States, began exploring the idea of using NMR to detect cancer. Damadian discovered that cancerous cells have different magnetic properties than normal cells, which could be detected using NMR technology.

He proposed that NMR could be used to detect and diagnose cancerous tissue in living patients. His research sparked the idea of creating detailed images of tissues based on their magnetic properties, but the process of turning this into a usable imaging technology was complex.

Development of the First MRI Scanners

While Damadian was one of the first to suggest that NMR could be used for medical imaging, it was Paul Lauterbur, an American chemist, who realized that by applying gradients to the magnetic field, NMR could produce two-dimensional images.

Lauterbur's breakthrough occurred in 1973 when he demonstrated that changing the magnetic field gradient could help reconstruct an image from the NMR signal. This allowed him to create the first NMR images of a test object, a significant advance over the one-dimensional signal originally produced by NMR.

In parallel, Dr. Peter Mansfield, a British physicist, made technical advances that significantly increased the speed of scanning and helped produce clearer images. Mansfield's contributions included a technique known as echo-planar imaging, which allowed for rapid imaging sequences.

The First Human MRI Scan and Damadian's Role

In 1977, Raymond Damadian built the first full-body MRI scanner, which he named "Indomitable." After several years of experimentation and refinements, Damadian and his team were able to create the first MRI scan of a human body.

Although Damadian's machine was groundbreaking, it was relatively slow and had limitations in image quality. Despite this, his early work paved the way for further refinements by other researchers and companies.

Commercialization and Development of MRI Machines

By the early 1980s, MRI technology was being developed commercially. Companies like General Electric, Philips, and Siemens improved the design and performance of MRI scanners, making them more practical and user-friendly.

These developments included faster scanning times, enhanced image resolution, and the ability to produce detailed cross-sectional images of different organs and tissues. Commercially produced MRI machines began appearing in hospitals, providing physicians with an unprecedented tool for non-invasive diagnosis.

Nobel Prize and Controversy

In 2003, Paul Lauterbur and Peter Mansfield received the **Nobel Prize in Physiology or Medicine** for their contributions to the invention of MRI. Lauterbur's work on imaging techniques and Mansfield's advances in rapid imaging were foundational in making MRI practical and efficient.

Raymond Damadian was notably not included in the Nobel Prize, despite his early contributions to MRI technology. This omission sparked controversy, as many felt Damadian's role in the invention was significant and deserving of recognition.

MRI Becomes Essential in Medicine

Throughout the 1980s and 1990s, MRI rapidly became an essential diagnostic tool for various medical fields. It allowed for detailed images of soft tissues, such as the brain, heart, muscles, and joints, which had previously been challenging to visualize with X-rays or CT scans.

MRI became especially valuable for detecting neurological disorders, assessing tumors, diagnosing musculoskeletal injuries, and identifying cardiovascular diseases.

Advances in MRI Technology

Over time, MRI technology has continued to evolve. Modern MRI machines use stronger magnetic fields, producing even higher resolution images and enabling specialized techniques such as functional MRI (fMRI), which can observe brain activity in real-time.

Advances in MRI have also included techniques like magnetic resonance angiography (MRA) for imaging blood vessels and diffusion MRI for studying neural pathways.

Significance of MRI

The invention of MRI was a pivotal moment in medicine, allowing for safer, more accurate diagnoses without invasive procedures or exposure to radiation. It represented a leap forward in how doctors could visualize and understand diseases within the human body. The collaboration between researchers in physics, chemistry, and medicine underscores the interdisciplinary nature of MRI's development and the profound impact of science on human health. Today, MRI continues to be a cornerstone of diagnostic imaging and research, benefiting millions of patients worldwide.

The story of the invention of the Email

The invention of email in the United States transformed communication, laying the foundation for the digital communication networks we use today. The development of email took place gradually, with contributions from several pioneering scientists and computer programmers. Here's how email came to be and the key innovations that led to its rise:

Early Computer Networks and Communication

In the 1960s, computers were large, expensive, and usually isolated from each other. Communication was limited to within individual systems.

The U.S. Department of Defense's Advanced Research Projects Agency (ARPA) developed ARPANET, one of the first computer networks, in the late 1960s. ARPANET allowed researchers to share resources and communicate across institutions by connecting computers over long distances.

Early messages on ARPANET were limited to basic commands and file transfers. While the concept of exchanging information was there, a true system for personal communication had yet to develop.

Ray Tomlinson and the First Email Program (1971)

The actual creation of email is credited to Ray Tomlinson, a computer engineer who worked for Bolt, Beranek, and Newman (BBN), a contractor for ARPANET.

In 1971, Tomlinson was experimenting with a file transfer program called CYPNET and an intra-computer messaging program called SNDMSG. He decided to combine these programs so messages could be sent from one computer to another over the ARPANET network.

Tomlinson also introduced the "@" symbol to separate the user's name from the computer's name in an email address, a convention that is still used today. His choice of "@" was largely due to its availability on keyboards and because it logically represented the phrase "at this address."

Early Email Usage on ARPANET

Tomlinson's program allowed users to send simple, text-based messages to each other across ARPANET. At first, it was used primarily by scientists and researchers for informal communication.

Email quickly became popular as it allowed users to send messages without having to be online at the same time, unlike phone calls or other real-time communication.

The functionality was basic, but it introduced core concepts of email, including sending, receiving, and addressing messages.

Development of Email Protocols and Features

As email gained popularity, ARPANET users saw the need for features like inboxes, folders, and "reply" options to organize messages. These developments came in the mid-1970s and were instrumental in transforming email into a more user-friendly system.

Protocols were also developed to standardize email communication. SMTP (Simple Mail Transfer Protocol), introduced in the early 1980s, enabled messages to be sent across different networks, expanding email's reach.

The creation of RFC 733 and later RFC 822 in the late 1970s and early 1980s provided guidelines for email structure and data format, facilitating compatibility between systems.

Email in Business and Public Use

By the early 1980s, email began to be adopted by businesses. Companies recognized that email could streamline internal communications, project management, and information sharing.

In 1982, CompuServe became one of the first companies to offer email services to its subscribers, making it accessible to a broader audience beyond academia and government.

Services like MCI Mail and America Online (AOL) in the mid-1980s and 1990s allowed everyday consumers to use email, introducing it to the general public and sparking rapid adoption.

The Growth of Internet Email and Modern Interfaces

As the internet expanded in the late 1980s and early 1990s, email services migrated from closed networks to the open internet, connecting users globally.

Tim Berners-Lee's invention of the World Wide Web in 1989 further popularized email, as more people gained access to the internet and could connect via websites.

Email services like Hotmail, Yahoo Mail, and later Gmail made email widely accessible and free. The web-based email interface, launched by Hotmail in 1996, allowed users to access email from any internet-connected device, contributing to email's widespread use.

Email's Role in Shaping Digital Communication

Over time, email became a staple of personal and professional communication. It replaced traditional mail for many purposes, allowing instant communication worldwide.

The impact of email has been profound in terms of both convenience and productivity. Email allowed individuals and organizations to connect across great distances, leading to the development of globalized business practices, faster information exchange, and remote collaboration.

Legacy of Email

The invention of email transformed the way people communicate, setting the stage for many of today's digital communication tools. It introduced the concept of "asynchronous" communication, where messages could be sent and received at different times. Email paved the way for other forms of digital messaging and remains a foundational part of internet infrastructure, used by billions of people globally for business, education, and personal communication.

Ray Tomlinson's innovation and the subsequent development of email protocols highlight how early internet technologies evolved into essential tools, underscoring email's lasting impact on both the digital world and daily life.

The story of the invention of the ATM

The invention of the ATM (Automated Teller Machine) in the USA revolutionized banking, providing 24/7 access to cash and enabling self-service transactions that transformed financial services. Although the concept of automated banking began internationally, the development and spread of ATMs in the United States were pivotal in making them a global norm. Here's the story behind how ATMs were introduced and popularized in the USA:

Early Automated Banking Concepts (1960s)

The idea of an automatic cash dispenser first surfaced in the 1960s. Luther George Simjian is often credited with creating a precursor to the ATM with his Bankograph, which he patented in 1960. It allowed customers to deposit cash and checks but was not widely adopted.

Around the same time, John Shepherd-Barron in the UK developed a cash dispenser that was first installed by Barclays Bank in 1967. This machine was one of the earliest examples of an ATM, but it required a paper voucher instead of a card and did not use PIN-based security.

The First ATMs in the USA: Don Wetzel's Breakthrough (1968)

In 1968, American engineer Don Wetzel, who worked for Docutel, an electronics company specializing in baggage handling equipment, had the vision to create a machine that would dispense cash. Wetzel came up with the idea after growing frustrated with long lines at bank tellers.

With support from Docutel, Wetzel led the development of the first true ATM. His machine allowed customers to withdraw cash using a plastic card and was the first to incorporate magnetic stripe technology, which stored account information.

In 1969, the first ATM in the United States was installed by Chemical Bank in Rockville Centre, New York. Chemical Bank launched the machine with the slogan, "On September 2, our banks will open at 9:00, and never close again."

Early Adoption and Growth of ATM Technology

The early ATMs only allowed cash withdrawals, and they required a special card provided by the bank. The machines could only connect to one specific bank's network, meaning that customers could only use their own bank's ATM.

Despite initial skepticism, these ATMs gained popularity due to their convenience, particularly among customers who couldn't always visit during bank hours. Banks saw the potential of ATMs to reduce teller costs and provide faster customer service.

Throughout the early 1970s, more banks in the USA began to adopt ATMs, but since each bank's machines were proprietary, customers still couldn't access cash from other banks' ATMs.

Development of Interbank Networks and PIN Security

The growth of ATMs led to the development of interbank networks to allow ATM access across different banks. In 1977, First National Bank of Atlanta launched the "MOST" network, which was one of the first regional ATM networks in the U.S.

The introduction of PIN (Personal Identification Number) codes also became a critical security feature, ensuring that customers could securely withdraw money from ATMs. This system was further refined and soon became the standard for ATM security.

Expansion and Evolution of ATM Networks (1980s)

In the 1980s, ATM networks expanded rapidly, with the formation of major national networks like Cirrus and Plus, which allowed ATMs across the country to link together. Customers could now access cash nationwide, regardless of which bank they used.

The ATM became a staple of American banking, with banks seeing it as both a convenience for customers and a way to operate more efficiently by reducing reliance on branch tellers.

Growth of ATM Features Beyond Cash Withdrawal

In addition to cash withdrawal, ATMs began offering other banking services, such as deposits, account balance inquiries, and transfers between accounts.

With advances in digital banking technology, ATMs continued to evolve, allowing customers to complete more complex transactions at any time, such as paying bills and obtaining cash advances on credit cards.

Global Impact and Worldwide ATM Adoption

By the mid-1980s, the concept of the ATM had spread globally, with ATMs appearing in Europe, Asia, and other parts of the world. As network interconnections grew, customers could access ATMs worldwide, transforming travel and international banking.

The ATM's global adoption demonstrated its appeal as a universal banking solution, enabling easy access to cash for people around the world.

Significance and Legacy of the ATM

The invention of the ATM changed banking permanently. By giving customers 24/7 access to their money and significantly reducing the need to visit a bank teller, ATMs made banking faster, more convenient, and accessible at any time of day. It was one of the first steps toward digital banking, foreshadowing today's online and mobile banking technologies.

The legacy of the ATM lies in its role as a precursor to modern digital banking, and it remains a crucial tool for accessing cash and performing banking transactions worldwide. Don Wetzel's contribution and the rapid expansion of ATM networks in the United States set a new standard for how people managed their money, bridging the gap between in-person banking and the digital future.

Around the World Series:

www.ingramcontent.com/pod-product-compliance
Lightning Source LLC
Chambersburg PA
CBHW082105220526
45472CB00009B/2049